T0213301

SpringerBriefs in Business

SpringerBriefs present concise summaries of cutting-edge research and practical applications across a wide spectrum of fields. Featuring compact volumes of 50 to 125 pages, the series covers a range of content from professional to academic. Typical topics might include:

- A timely report of state-of-the art analytical techniques
- A bridge between new research results, as published in journal articles, and a contextual literature review
- A snapshot of a hot or emerging topic
- An in-depth case study or clinical example
- A presentation of core concepts that students must understand in order to make independent contributions

SpringerBriefs in Business showcase emerging theory, empirical research, and practical application in management, finance, entrepreneurship, marketing, operations research, and related fields, from a global author community.

Briefs are characterized by fast, global electronic dissemination, standard publishing contracts, standardized manuscript preparation and formatting guidelines, and expedited production schedules.

More information about this series at http://www.springer.com/series/8860

James A. Cunningham • Brian Harney •
Ciara Fitzgerald

Effective Technology Transfer Offices

A Business Model Framework

 Springer

James A. Cunningham
Newcastle Business School
Northumbria University
Newcastle upon Tyne, UK

Brian Harney
DCU Business School
Dublin City University
Dublin, Ireland

Ciara Fitzgerald
Cork University Business School
University College Cork
Cork, Ireland

ISSN 2191-5482　　　　　　　　ISSN 2191-5490　(electronic)
SpringerBriefs in Business
ISBN 978-3-030-41944-8　　　　ISBN 978-3-030-41946-2　(eBook)
https://doi.org/10.1007/978-3-030-41946-2

This Springer imprint is published by the registered company Springer Nature Switzerland AG.
The registered company address is: Gewerbestrasse 11, 6330 Cham, Switzerland

*To my loving family Sammi, Aiden and Riley
and to a wonderful great granddad Derek
(JC)*

Preface

Historically there has been a failure to capitalise on the true value of university business collaboration and thus realise the potential that benefits business and society. Successful universities in the twenty-first century need to excel in teaching and research as well as in developing complementary third stream activities to exploit research for private and public good outcomes. With public funding for science becoming more demanding in terms of demonstrating impact, this in turn impacts researchers and universities. The role of technology transfer offices (TTOs) has become even more strategic in ensuring that more technology transfer activities are delivered for less for private and public outcomes. TTOs are now regarded as key actors in regional and national innovation systems. Given the multiple roles of TTOs as information brokers, science marketers and catalyst for academic entrepreneurship, TTOs provide a crucial mechanism to allow universities respond to this challenge. However, in spite of the increased emphasis placed on the TTO to deliver, little is known about the 'black box' of the TTO (Sanders and Miller 2010).

While international evidence from best practice provides a basis for understanding the requirements for success, individual universities will have to find their own solutions to these issues, depending on their size, disciplinary base and the economy of their region. A key challenge for universities is how best to organise and differentiate their third mission activities that ensures sustainable, effective and impact-orientated technology transfer? University research commercialisation has grown and developed in the USA since the Bayh Dole Act in 1980. Many countries and universities in established and emerging markets have imitated many aspects of US university technology transfer offices activities, structures and processes. TTOs worldwide are looking to emulate the blockbuster technology success of American universities. The growth of TTOs and personnel has been the subject of growing empirical investigation and studies. However, a core challenge remains for TTOs and practitioners by way of accessible tools and frameworks that allows them to understand the key success factors and the strategic issues to become effective with respect to effective technology transfer that combines key strategic and operational issues.

In developing accessible frameworks for TTOs, the core understanding is that a 'variety of excellence' exists and allows each technology transfer office and university to choose its own distinctive strategy. Therefore, an overarching business model framework for TTOs is appropriate for practitioners which recognises the varieties of excellence but also one that draws on elements of best practice elsewhere, acknowledging that policy choice and implementation will have to be adapted to suit local circumstances and contingencies.

Combining elements of best practices, previous empirical studies and our own research on TTOs, ecosystems and scientists in the principal investigator role, we have developed a business model framework for TTOs, which captures key elements of their strategic and operational activities that need effective management and leadership from a practitioner perspective.

Our business model framework for TTOs addresses strategic issues including strategy, organisational structure, staff and resources, activities, mechanisms, policy and procedures and evaluation and outcomes. The business model framework considers contextual factors that directly and indirectly impact on TTOs' commercialisation culture and ethos and researcher commitment, awareness and motivation. For each element of our business model framework the core key success factor and the facilitating factors that enable effective technology transfer with TTOs are outlined.

Newcastle upon Tyne, UK   James A. Cunningham
Dublin, Ireland   Brian Harney
Cork, Ireland   Ciara Fitzgerald

Reference

Sanders CB, Miller FA (2010) Reframing norms: boundary maintenance and partial accommodations in the work of academic technology transfer. Sci Public Policy 37(9):689–701

Acknowledgements

We wish to acknowledge the support of Brian O'Kane from Oak Tree Press in Ireland for allowing us to use elements of from Cunningham, J. A. and Harney, B. (2006) Strategic Management of Technology Transfer: A New Challenge on Campus. We would like to acknowledge the support of TTO professionals in Ireland, the UK and the USA in further shaping our ideas and thoughts. James A. Cunningham and Ciara Fitzgerald wish to acknowledge the funding support from the PRTLI 4 Irish Social Science Platform. James A. Cunningham wishes to acknowledge the funding support of Science Foundation Ireland (SFI) and co-funding under the European Research Development Fund under Grant Number 13/RC/2073. Finally, from Springer we would like to thank Prashanth Mahagaonkar and Srinivasan Manavalan for their editorial support and patience.

Contents

About the Authors

James A. Cunningham is a Professor of Strategic Management at Newcastle Business School, Northumbria University. He previously held academic positions at University College Dublin and the National University of Ireland Galway (NUI Galway). At NUI Galway he held a range of leadership positions including Head of Strategic Management group, Executive MBA Programme Director, Director of the Centre of Innovation and Structural Change and founding Director of the Whitaker Institute. His research intersects the fields of strategic management, innovation and entrepreneurship. His research focuses on strategy issues with respect to scientists as principal investigators, university technology transfer commercialisation, academic, public sector and technology entrepreneurship, entrepreneurial universities and business failure. He has papers published in leading international journals such as *Research Policy, Small Business Economics, R&D Management, Long Range Planning, Journal of Small Business Management* and *Journal of Technology Transfer*, among others. Awards for his research include six best paper conference awards and two case study international competition awards. He has published several books and research monographs on the themes of strategy, entrepreneurship, technology transfer and technology entrepreneurship with leading publishers such as *Oxford University Press, Palgrave MacMillan, Springer* and *World Scientific Publishing*. He has been an editor of special issues for *Journal of Management Development, Journal of Technology Transfer, Journal of International Entrepreneurship, Irish Journal of Management* and *Small Enterprise Research*. He has delivered keynote talks and presentations nationally and internationally to business, policy and academic audiences, as well as delivering executive master classes on strategy development, innovation and technology entrepreneurship. He also has developed and delivered *Positioning Your Research for Impact* workshops for postdoctoral and early career researchers.

Brian Harney is Associate Professor in Strategy and HRM at Dublin City University Business School and former Programme Director of the award-winning MSc in HRM (gradIreland Best Postgraduate Course 2016). In 2019, he was Chair of the

AOM HR Division International Conference hosted by DCU. Brian's research focuses on the intersection of Strategy and HRM, with a particular focus on SMEs, growth and knowledge-intensive sectors. His research has received over 15 Best Paper awards and is published in *Human Resource Management Journal*, *International Journal of Human Resource Management* and *Advances in Industrial and Labor Relations*. Brian is the co-author of three books including *Strategy and Strategists* (Oxford University Press) and an edited collection *Strategic HRM: Research and Practice in Ireland* (Blackhall/Orpen). He is the only DCU faculty member to have won both categories of the President's Award for Excellence in Teaching and Learning (new lecturer and overall award). He has held visiting positions in China, Slovenia, Malta and the USA, most recently at the University of Southern California. Brian serves on the Editorial Board of HRM (USA), the *International Journal of HRM*, *Employee Relations* and the *Irish Journal of Management*. He is currently DCU Principal Investigator of the EU RISE funded GETM3 project and academic lead for the Go Global 4 Growth Management Development programme and serves as a Judge for the HR Leadership Awards (Ireland) and the annual Gov HR Summit (GCC). In 2012, Brian delivered a 24-hour non-stop lecture for charity and was nominated for a President's Award for Civic Engagement.

Ciara Fitzgerald is a faculty member at the Cork University Business School, University College Cork. Her research and teaching interests lie at the intersection of technology innovation, entrepreneurship and strategy. Her research investigates (1) strategies used by universities and firms to manage intellectual property and the commercialisation process, (2) strategies to engage citizens and policymakers in technology assessment and (3) exploratory and applied research of innovative health information systems. Ciara has published widely in leading journals and influential volumes such as *Research Policy*, *Journal of Technology Transfer*, Edward Elgar, Routledge and Cambridge University Press. Prior to joining UCC as a faculty member, Ciara completed a senior postdoctoral fellowship and a postdoctoral fellowship at the Smurfit School of Business, University College Dublin. She was awarded her PhD from the National University of Ireland Galway. Ciara's industry experience includes consulting experience at Accenture. During her time there she gained valuable experience working on a multitude of projects for clients in the financial services sector. She is Co-Director of the Health Information Systems Research Centre in the Department of Business Information Systems in UCC.

Chapter 1
Technology Transfer Offices: Roles, Activities, and Responsibilities

Abstract Technology transfer offices' roles, activities, and responsibilities form the basis for an effective technology transfer business model framework. Taking a historical context TTOs main roles were namely switchboard services, network development, and technology transfer and managing IP activities. However, the TTO role, activities and responsibilities have expanded to meet increasing internal and external pressures. To this end, we outline some of the multiple responsibilities of TTOs and conclude by exploring the development of expertise that is now required.

Keywords TTOs · Technology transfer · Skills · Activities · Roles · Responsibilities · Mission statements

1.1 Introduction

The last decades have seen a significant growth in universities setting up Technology Transfer Offices (TTOs). This is driven by a myriad of factors, especially as policymakers now view universities as key and pivotal actors within entrepreneurial and innovation ecosystems. In this chapter, we firstly focus on the roles and activities of TTOs. We then turn our attention to focus on the key responsibilities of TTOs before concluding the chapter by exploring the different facets required for TTOs to develop their expertise.

1.2 Roles and Activities of TTOs

TTOs traditionally play several important roles in the process of technology transfer including information broker, science marketer, and catalyst for academic entrepreneurs (Fassin 2000). Several leading international research universities, including the University of California, Stanford, MIT, and the University of Wisconsin, established their respective Technology Transfer Offices over 40 years ago. The

Research Corporation, founded in 1912 by a University of California-Berkeley faculty member to sell rights to use patents of several affiliated universities, was a predecessor to these contemporary offices (Mowery et al. 2001). In essence as Kreiling and Bounfour (2019: 1) put it 'Technology Transfer Organisations (TTO) are important at the intersection between academia and industry'.

Definitions of the various roles that TTOs should pursue diverge between those who favour a narrow role for TTOs—primarily as a switchboard—and those who favour a broader role—of helping two-way communications between HEIs and the outside world such as identifying curriculum development needs. Yet this is a simplistic dichotomy, in reality the function of TTOs varies from HEI to HEI and from country to country. Overall the main function of a TTO is to provide a formal, above the board, and a relatively effective mechanism for those researchers who wish to commercialise their ideas. Göktepe-Hultén (2010: 41): 'TTOs make it possible to scientists to transfer technology through different mechanisms and through different agents' The role of TTOs can be broken down into four main areas, namely *switchboard services, network development, management of technology transfer, and managing IP activities.*

Switchboard Services

Technology transfer as a formal function of HEIs involves the management of the interface between academia and various external institutions (see Schaettgen and Werp 1996). HEIs are often organisationally complicated institutions, and businesses can find it very difficult to access relevant expertises and resources. In particular, SMEs can be put off if there is no obvious point of entry to the HEI's resources. This is the traditional role of the TTO, serving as a signpost offering switchboard services and directing industrialists who are seeking the most appropriate expertise within the HEI (Jones-Evans 1998). A key role of the TTO involves promoting HEI–industry interaction. The goal of the office is to lower the entrance barrier for the external business world and to complement existing informal direct contacts between faculty and industrial representatives.

Network Development

HEIs can play a proactive role in developing strong linkages with industry, particularly through the TTO, which has been highlighted (Scott et al. 2001). TTOs facilitate this through providing opportunities for enterprising staff to develop ideas and contribute to the economy via knowledge transfer and spin-off companies. Best practice examples of TTO activities are largely US-based examples where TTOs have extensive resources and staff to facilitate commercialisation and HEI support of these activities is taken as a given (e.g. MIT, University of California). The reality in most other countries highlighted by Jones-Evans et al. (1999: 49) suggest: 'little information to suggest that ILOs are undertaking a pro-active role..., at most, industrial liaison offices at higher educational institutions are merely providing marketing services for their parent organisation'. Moreover, Jones-Evans et al. (1999) conducted a comparative study of TTO roles and activities in Ireland and Sweden, using interviews with key individuals involved in the process of technology transfer in both countries. Their results indicated that the role of TTOs should be to provide a strategic focus for HEI–business collaboration although how this became manifest differed by country. In Sweden, the TTO provides a gateway to a network of technology transfer organisations. In Ireland, the development of both internal and external formal networks was judged to be in a primitive state. Subsequently, the university-based technology transfer system has changed considerably with the creation of Knowledge Transfer Ireland.

An inherent part of TTO activities involves developing soft skills and maintaining contacts. Such relationship building is an ongoing process, which is crucial to the development of TTOs and in creating further opportunities for the commercialisation of research (Meyer 2003). Research by Siegel et al. (2003) highlight the importance of such informal TTO activities and noted: 'relationship building, personal relationships were mentioned more frequently than contractual contacts, emphasising the continual important informal aspect of communication and relationship building to the technology transfer process'. Such findings indicate that a key role for TTOs should be the formation of networks, which as all stakeholders emphasised are important in technology transfer. Sanchez and Tejedor (1995) on the establishment of formal links between TTOs and business highlighted a number of key mechanisms, which include:

- Firm managers themselves search for university departments
- Managers receive proposals from university departments
- Managers seek support from TTO
- Managers receive proposals from TTO

Management of Technology Transfer
In general, the function of the TTO is to manage the activities that occur as part of the collaborative efforts between HEIs and industry. The TTO manages the process of invention disclosure and evaluation, takes decisions on what inventions to patent and facilitates with the negotiation of licencing agreements. As a result of this commercialisation role, the TTO has direct and frequent contact with industry. This contact can often spawn indirect effects such as student placement and general support for the university. Although technically a step removed from the actual research, the TTO should have familiarity and good contacts with the expertise of faculty and track potential industrial research opportunities. By conducting these activities the TTO plays a central role in support of the HEI. However, as Fitzgerald and Cunningham (2016: 1236) highlight: 'At the early stages of establishing TTOs defining a clear purpose is critical to establish legitimacy and credibility'. How TTOs do this will vary based on the needs of local business and society, the mission of the university, and the focus of the local economy.

Management of Intellectual Property
A primary role that TTOs are involved in is the management of intellectual property (IP). The share of IP activities managed by the TTOs for the host institution varies across OECD countries as highlighted by the OECD (2003: 41). 'The degree to which IP activities are concentrated in a TTO or administrative unit of the PRO tells much more about the TTO's formal organisation and capacity'. For example, in Japan 35% of PRO's claim to manage 100% of IP for their institution, while 23%

claim to manage between 25% and 75%. This reflects the use of separate or independent TTOs to manage IP for national universities. Some 34% of Swiss university-based TTOs reported that they manage up to 100% of the IP of their institution. Thus, the extent to which IP is managed centrally by a TTO is contingent on IP policies and the underlying structure of the Technology Transfer Office. In their UK study, Meyer and Tang (2007) found how UK universities managed IP varied considerably. In order to ensure the TTOs maximise the value of the IP that they manage Holgersson and Aaboen (2019) argue that they should move from an appropriation to a utilization model.

1.2.1 A Typology of TTO Roles and Associated Activities

Combining research (see Allan 2001; Baglieri et al. 2018; Jones-Evans et al. 1999; Fassin 2000: 38; Harman and Stone 2006; Lambert 2003: Miller et al. 2018; O'Kane 2018; Scott et al. 2001; Sideri and Panagopoulos 2018; Siegel et al. 2003), secondary source data along with primary interviews that we have done over the years we have developed a typology of TTO roles and associated activities. This model seeks to operationalise and builds on Fassin's (2000) three technology transfer roles—information broker, science marketer, and catalyst for academic entrepreneurs by relating them to activities conducted by TTOs (see Table 1.1). From this, it is evident that TTOs fulfils many different roles internally and externally.

1.3 Responsibilities of TTOs

Universities with large research budgets and a 'higher academic excellence' base are typically first adoptors of creating TTOs (see Castillo et al. 2018). Irrespective of the institutional adoption timing for TTOs, in essence, the ultimate responsibility of TTOs is '*guardian of the university's intellectual property*' (Fassin 2000: 37). Yet this is narrow in its depiction as often TTO's responsibilities extend beyond the mere protection of formal IP (see Knowledge Commercialisation of Australasia 2003). As noted by the OECD (2003: 59): 'TTOs do far more than simply ensure the protection of patentable inventions. They are often involved in protecting and exploiting innovations in a number of technological fields'.

TTOs often receive invention disclosures and are responsible for deciding which innovations and creative works to protect and how to do so. They may also be involved in negotiating contracts that stipulate how IP is to be used or revealed. In this role, the responsibility of the TTO is not to get the patent but rather *to close the deal*. The TTO's priority, therefore, is getting the technologies into the marketplace for the public good (Meyer 2003: 11). Secondary responsibilities include promoting technological diffusion and securing additional research funding for HEIs via

Table 1.1 Core roles and activities of TTOs

Role	Task	Illustrative activities
Information broker	Dissemination of information	• Fielding inquiries from industry • Development and distribution of marketing material online and offline • Promoting networking opportunities • Acting as an information point • Commercial advice
Science marketer	Marketing and promotional activities	• Organising visits to laboratories • Participating in conferences and presentations • Participating in specialised technology fairs
Science marketer	Public relations activities	• Networking with professional associations • Writing articles in periodicals and press to support technology transfer activities • Promoting special events • Marketing the university and activities of the technology transfer office
Catalyst for academic entrepreneurs	Intellectual advice and negotiation	• Advising on intellectual property and patent issues • Negotiating agreements • Licencing • Defining strategy for technology transfer
Information broker/science marketer/catalyst for academic entrepreneurs	Active management and valorisation of university potential (technology transfer)	• IP management-identify/evaluate/protect • Searching for industrial and commercial partners • Searching for financial partners (venture capital funds, business angels) • Supporting spin-off companies • Supporting business incubators and accelerators
Information broker	Aggregating data	• Building information systems and analytical methodologies to support partner searches • Maintaining/supporting of directories of research activities and expertise
Information broker/science marketer/catalyst for academic entrepreneurs	Coordination	• Of the university research park • Of the university incubator centre • Of the university support science park and or accelerator programmes • Of the university see capital fund

Table 1.2 Sample mission and vision statements: the responsibilities and roles of TTOs

University	TTO mission and visions
Warwick University (1)	Warwick Ventures commercialises innovations produced from world-leading research at the University of Warwick. We offer advice and services to the University's innovators. Our role is to support them throughout the process of generating impact and a commercial return from their research, whilst they maintain their academic focus. We work closely with industry. Our specialist commercialisation managers provide businesses and investors access to the best of the University's Intellectual Property. We support technology development, licence IP, and create spin-out companies that retain close ties to the University's expertise. Our innovations span a wide range of technologies and sectors, and new innovations emerge every week.
Columbia University—Columbia Technology Ventures (2)	CTV's core mission is to facilitate the transfer of inventions from academic research labs to the market for the benefit of society on a local, national, and global basis.
Trinity College Dublin Office of Corporate Partnership and Knowledge Exchange (3)	The OCPKE *supports both industry engagement and the commercialisation of Trinity research*. The office reaches out to industry and the business community to develop partnerships that enable the industry to *benefit from the world-leading teaching, research, and infrastructure within Trinity*.
Northwestern University Innovation and New Ventures (4)	INVO catalyzes the translation of Northwestern innovations to benefit the public and promote economic growth. In order to maximize that outcome, Northwestern follows important principles when licencing university technology.
MIT Technology Licensing Office (5)	We cultivate an inclusive environment of scientific and entrepreneurial excellence and bridge connections from MIT's research community to industry and startups by strategically evaluating, protecting, and licencing technology.
The University of Hong Kong Technology Transfer Office (6)	Our Vision is to be the Leading Innovation and Enterprise Partner. We strive to create a positive impact through innovation based on the research of HKU for the benefits of communities and the advancement of technologies. We also organize educational events to foster an innovative and entrepreneurial culture.

Source: (1) https://warwick.ac.uk/services/ventures/about/ (accessed 1 November 2019); (2) https://techventures.columbia.edu/about-ctv/technology-transfer-columbia; (3) https://www.tcd.ie/innovation/industry/; (4) https://www.invo.northwestern.edu/about/mission/index.html; (5) https://tlo.mit.edu; (6) https://www.tto.hku.hk/about-us

royalties, licencing fees, and sponsored research agreements. One of the main responsibilities of TTOs is to assess and protect IP and make it available for public use (OECD 2001, 2003). This is evident from the sample mission statements in Table 1.2, which depict the key responsibilities and roles of a number of TTOs.

As TTOs office develop their institutional remit can expand to beyond protecting university technology transfer. Fitzgerald and Cunningham (2016) in their study of Irish TTO mission statements found their mission statements predominately focused externally on their customers and markets and also the activities that they offer and they conclude by noting: 'Mission statements do matter for TTOs. Mission statements are important artifacts for TTO Directors and TTOs in communicating effectively who they are and what do they do to any stakeholder and audience' (Fitzgerald and Cunningham 2016: 1244). In an intercountry TTO comparison study by Jefferson et al. (2017) found that: 'Societal benefit and support for research and central objectives in the mission statements of all five of the TTOs studied'.

1.3.1 TTO Responsibilities to Multiple Stakeholders

TTOs are subject to a number of competing demands and have to meet the needs of various stakeholders, including faculty, administration, commercial, and state priorities (Graff et al. 2002). These multiple responsibilities are particularly evident in some activities, e.g. the provision of legal and intellectual property management services to researchers and the collection of licencing royalty revenues for the HEI from the industry. The revenues generated by licencing, while still only a minor percentage of HEIs' operating budgets, have grown substantially. Beyond their monetary value, the growth of such royalty revenues serves to demonstrate the success of TTOs in diffusing the fruit of the HEI's research. In a study undertaken to review the economic effects of the Bayh Dole Act, Jensen and Thursby (2001) surveyed 62 research universities concerning their technology transfer activities. Technology Transfer Offices, faculty, and the university administrators were all interviewed about their objectives for technology transfer. The different parties' responses demonstrate the differences that exist within the HEIs over technology transfer. TTO Officers prioritised revenue whereas faculty members ranked sponsored research as their number of ranked priority. Administration, discussions with academics, and people outside the university where the top three activities that the Australian study of technology transfer managers devoted considerable hours per week (Harman and Stone 2006).

Several studies have examined how much time that faculty allocate to technology transfer activities and this is dependent such factors as the type of research grants, commercial experience, and the support they receive from their host institution to purse technology transfer (see Arvanitis et al. 2008; Cunningham et al. 2016; Libaers 2012; Link et al. 2017). However, university-based administrators tend to consider technology-associated revenues most important, while the faculty see the ability to attract research sponsorship as paramount. TTOs often operate under somewhat conflicting mandates from their administration and faculty and emphasise the more immediate and tangible outcomes of executed licences and commercialised inventions.

1.4   Developing Expertise

A critical dimension for an effective technology transfer business model is having the appropriate balance and skill mix set of technology transfer staff. Investing in staff to achieve the optimal mix in terms of experience and skill sets is essential. The extent of developing such expertise in universities is subject to resource constraints and the strategic priority that university administrators place on technology transfer activities as well as the wider institutional innovation and entrepreneurship agenda.

1.4.1   Levels of Staffing

There have been different studies that have highlighted the staffing levels of TTOs and technology transfer professionals. Surveys of OECD countries show that TTOs over two decades ago had fewer than five full-time staff (2003: 12). Moreover, the same study showed that for most countries the staff numbers at TTOs are relatively low but are growing. In Norway, for example only 1 out of 5 of survey respondents have more than one full-time equivalent (FTE) dedicated to technology transfer issues. In the United States, the number of TTO staff is somewhat larger with a mean of 3.3 staff devoted to licencing issues (Table 1.3).

For example, in the mid-2000s in an Irish context, not all HEIs had clearly identified the role and function of IP management or had created specific positions to manage it. Those institutions that had approximately 62 people (22 full-time equivalents (FTE)) were regarded as being involved in commercialisation activities—providing an average of 0.96 FTE staff per research institution. This compared with a median of 2.2 FTE licencing staff and 1.8 FTE other staff at the TTO offices in the United States at the time. Resources at TTO offices within Irish TTOs were thus deemed inadequate to ensure successful commercialisation of research (Forfas 2004). However, with the creation of Knowledge Transfer Ireland and supported by different strengthening initiatives this meant a significant transformation of the Irish technology transfer system (see Enterprise Ireland 2016). The first of these strengthening initiatives (TTSI1) directly funded 32 technology transfer professionals across then third-level institutions (see Enterprise Ireland 2014).

Table 1.3   Employment at TTOs at US universities (full-time equivalent), 2000

	Licencing staff in FTE	Other staff
Total	562.5	586.5
Mean	3.3	3.5
Median	2.0	1.8
Number of responding universities	168	168

Source: US Technology Administration, Department of Commerce, based on AUTM 2000 data

1.4.2 Technology Transfer Skills

In the United States, the land grant universities (i.e. those that have been granted land upon which to build in return for delivering services back to the local community) have had the application of knowledge at the heart of their mission from their inception. The contrasts with the situation in many EU countries where commercialisation activities are not historically supported and grounded (PACEC 2003). However, in the last decade with various national initiatives among EU member states significant progress has been with developing technology transfer skills among stakeholder supporting technology transfer. An example of such initiatives is Progress TT Capacity Building for Technology Transfer funded under Horizon 2020.[1]

Technology transfer is people intensive and requires a wide and specialised set of skills. Many HEIs face problems in building professional offices on their own. Protecting and managing IP requires specific legal knowledge, along with commercial and scientific expertise. For example, licencing needs a combination of market awareness, subject-specific knowledge, marketing, and negotiating skills. Spin-out creation requires entrepreneurship skills, links with business angels and venture capitalists, business planning, management, and company formation expertise. These skills are difficult to find in a small group of people and are expensive to buy-in.

The significant expansion in the roles, activities, and responsibilities of TTOs has meant that an increasingly more diverse and advanced skill set is needed. These demands are further complicated by multiple stakeholder influence. The range of skills required by TTOs as identified by Shattock (2001) includes:

- The ability to build networks
- A capacity for brokerage
- A wider vision about the university and the economy
- Ability to marrying market niches and gaps
- Strategic skills in identifying university research strengths
- Legal and intellectual property skills
- Skills in company formation

1.4.3 Personnel Profile

Some of TTOs in the United States place a strong emphasis on recruiting staff with substantial industry experience, and find it difficult to teach the negotiation and deal-making skills learnt in industry to new staff. The tendency traditionally in Europe is that TTOs are staffed by academics or university administrators but this has begun to

---

[1]http://www.progresstt.eu

change. This can create barriers to the negotiation of contracts as business generally finds it easier to negotiate with individuals who have more of a commercial background. Yet a major restriction in addressing this deficiency is the limited salary to attract experienced entrepreneurs or industry executives into TTOs. The Lambert Report (2003) noted that UK advertisements still focus on the subject background, rather than functional experience. In examining the key background demographics of technology transfer managers based on Australian data Harman and Stone (2006) found that 60% were males and 40% were female, 31% held a doctorate, 26% held an MBA, and 16% held a law qualification. Allan's (2001) study of best practice found that TTOs noted difficulties in training faculty in patenting and the innovation process. Of the institutions that were studied, TTOs with longstanding operations seemed to draw more from years of precedence and reputation to maintain and enhance the technology transfer process (Allan 2001). EIMS (1995) noted, however, that often researchers will often have industrial experience: 'some of the most successful technology transfer programmes are invariably initiated and taken care of by senior researchers who both excel academically in their specialist discipline and have solid industrial experience or have held senior positions in related industries'. Having capable staff is crucial as noted by the OECD (2003: 46): 'well trained staff at TTOs are not only essential to the efficiency of technology transfer but can also help limit conflicts of interests with researchers'. Training needs to be both formal and informal, and should address the skill deficits in the following areas:

• Appropriate research protocols
• Logging and recording of research findings
• The patent process, patent protection, patent law
• The costs of commercialisation

1.4.4   Functions of TTO Staff

According to Allan's (2001) best practice study found that most TTOs stress the importance of informal relations and relation building with academics. This may suggest why in many cases it is recommended that staff hired for TTOs have industry experience and hence an understanding of industry as well as a network of contacts and linkages (OECD 2003). A study by Balthasar et al. (2000) found: 'successful institutions at the interface between science and industry do not consider themselves to be an institution for transfer but a network manager'. TTOs need to build effective relationships with academics and business and they must ensure that they have a distinctive identity that supports their legitimacy (see O'Kane et al. 2015). Furthermore, O'Kane (2018) found that TTOs have a more backward integration focus through increasing their attention researchers as they have capabilities challenges in relation to market-facing activities. In their content analysis of 110 job descriptions of TTO directors mainly covering the Anglo-Saxon context, i.e. Australia, Ireland, the United Kingdom, and the United States, Cunningham and Menter (2019) found

Table 1.4   Functions and activities of Technology Transfer Managers/Directors

• Develop and implement an intellectual property and technology commercialization strategy
• Contribute to the development of institutional policies relating to licencing of intellectual property
• Responsible for institutional patent prosecution and management strategy
• Screen technological change and market evolutions
• Identify and evaluate inventions generated by researchers
• Adopt technology transfer options to meet continuously changing demand
• Search for new clients
• Define collaborative projects
• Ensure professional project management
• Manage patent filing and patent prosecution
• Negotiate confidentiality, option, licencing, sponsored research, and collaboration agreements
• Managing working relationships with industrial clients and researchers
• Facilitate new company formation
• Counsel faculty, staff, and students in matters relating to intellectual property, licencing, and conflicts of interest
• Deliver outreach programs to faculty, staff, and students

the job requirement profiles focused on education, work experience, negotiations, IP protection, collaboration, and strategy. The TTO director job description mainly focused on IP protection, operations, and internal and external collaborations. The functions and activities carried out by Technology Transfer Managers/Directors are multiple, as illustrated in Table 1.4.[2] In essence, the functions of TTOs are complex and diverse with their performance outcomes wholly reliant on other stakeholders.

1.5   Implications for TTOs and University

The implications for the universities are clear. In order to develop technology transfer initiatives, university leaders and administrators have to have a clear strategic focus and plan for such activities. In practice, this means developing a strategic plan which is cognisant of the current status of such activities within their own institutions, as well as further aspirations that are aligned to different stakeholder requirements. The institutional environment and the needs of the region in which the university is situation does matter. Moreover, university leaders need to ensure the TTO can be integrated and scaled accordingly aligned to the current and future needs and demands. Consequently, by clearly defining roles, activities and responsibilities of TTOs, directors of such offices can manage technology transfer activities effectively and become more than a guardian and champion of intellectual property. Specifically based on this chapter some key issues can be extrapolated:

---

[2]Based on EIMS (1995), Scott et al. (2001) and sample job descriptions from Georgetown University provided by Dr. Martin Mullins and other secondary source data.

- Each TTO should have a clear mission statement and strategic plan, where key roles, activities, and responsibilities are clearly defined.
- Technology Transfer activities involve a mix of roles, including being an information broker, a science marketer, and a catalyst for academic Entrepreneurs.
- A core responsibility of TTOs is as the guardian of the university intellectual property.
- Stakeholders of TTOs have different motives and incentives. Consequently, one of the main challenges that TTOs face is managing expectations and communicating with the stakeholders in a manner that potentially diffuses or helps to resolve these differences.
- Soft skills in technology transfer are crucial to its development and for creating further opportunities for the commercialisation of research.
- TTOs place significant efforts in training faculty in patenting and in the innovation process.
- The rapid growth of TTOs' roles and activities requires them to develop a range of skills including:

  - The ability to build networks
  - A capacity for IP brokerage
  - A wider vision about the university and the economy
  - Ability to marrying market niches and gaps
  - Strategic skills in identifying university research strengths
  - Legal and intellectual property skills
  - Skills in company formation

- Successful TTOs place a strong emphasis on recruiting staff with substantial industry experience and invest in staff development.

References

Allan MF (2001) A review of best practices in university technology licensing offices. J Assoc Univ Technol Manag 13(1):57–69

Arvanitis S, Kubli U, Woerter M (2008) University-industry knowledge and technology transfer in Switzerland: what university scientists think about co-operation with private enterprises. Res Policy 37(10):1865–1883

Baglieri D, Baldi F, Tucci CL (2018) University technology transfer office business models: one size does not fit all. Technovation 76:51–63

Balthasar A, Battig C, Thierstein A, Wilhelm B (2000) 'Developers': key actors of the innovation process. Types of developers and their contacts to institutions involved in research and development, continuing education and training, and the transfer of technology. Technovation 20 (10):523–538

Castillo F, Gilless JK, Heiman A, Zilberman D (2018) Time of adoption and intensity of technology transfer: an institutional analysis of offices of technology transfer in the United States. J Technol Transf 43(1):120–138

Cunningham JA, Menter M (2019) Technology transfer office directors: an exploratory study of job roles, responsibilities and impact. R&D Management Conference, Paris

Cunningham JA, O'Reilly P, Dolan B, O'Kane C, Mangematin V (2016) Publicly funded principal investigators allocation of time for public sector entrepreneurship activities. Econ Polit Ind 43 (4):383–408

EIMS (1995) Good practice in the transfer of university technology to industry: case studies by a consortium led by inno GmbH, publication no 26. European Innovation Monitoring System, Brussels

Enterprise Ireland (2014) A review of the performance of the Irish technology transfer system 2007–2012. https://www.knowledgetransferireland.com/Reports-Publications/A-review-of-the-performance-of-the-Irish-technology-transfer-system-2007-2012.pdf

Enterprise Ireland (2016) A review of the Irish technology transfer system 2013–2016. https://www.knowledgetransferireland.com/Reports-Publications/A-Review-of-the-Performance-of-the-Irish-Technology-Transfer-System-2013-20161.pdf

Fassin Y (2000) The strategic role of university-industry liaison offices. J Res Adm 1(2):31–42

Fitzgerald C, Cunningham JA (2016) Inside the university technology transfer office: mission statement analysis. J Technol Transf 41(5):1235–1246

Forfas (2004) From research to the marketplace: patent registration and technology transfer in Ireland. Forfas, Dublin

Göktepe-Hultén D (2010) University-industry technology transfer: who needs TTOs? Int J Technol Transf Commer 9(1):40

Graff G, Heiman A, Zilberman D, Castillo F, Parker D (2002) Universities, technology transfer and industrial R&D. In: Evenson RE, Santianello V, Zilberman D (eds) Economic and social issues in agricultural biotechnology. CABI Publishing, New York, pp 93–117

Harman G, Stone C (2006) Australian university technology transfer managers: backgrounds, work roles, specialist skills and perceptions. J High Educ Policy Manag 28(3):213–230

Holgersson M, Aaboen L (2019) A literature review of intellectual property management in technology transfer offices: from appropriation to utilization. Technol Soc 59:101132

Jefferson DJ, Maida M, Farkas A, Alandete-Saez M, Bennett AB (2017) Technology transfer in the Americas: common and divergent practices among major research universities and public sector institutions. J Technol Transf 42(6):1307–1333

Jensen R, Thursby M (2001) Proofs and prototypes for sale: the licensing of university inventions. Am Econ Rev 91(1):240 259

Jones-Evans D (1998) Universities, technology transfer and spin-off activities: academic entrepreneurship in different European regions. University of Glamorgan Business School, Glamorgan

Jones-Evans D, Klofsten M, Andersson E, Pandya D (1999) Creating a bridge between university and industry in small European countries: the role of the industrial liaison office. R&D Manag 29(1):47–56

Knowledge Commercialization of Australasia (2003) Forum and fair ideas: commercialisation. Discussion paper. Canberra: Knowledge Commercialization of Australasia

Kreiling L, Bounfour A (2019) A practice-based maturity model for holistic TTO performance management: development and initial use. J Technol Transf:1–30

Lambert R (2003) Lambert review of business-university collaboration. https://papers.ssrn.com/sol3/papers.cfm?abstract_id=1509981

Libaers DP (2012) Time allocation decisions of academic scientists and their impact on technology commercialization. IEEE Trans Eng Manag 59(4):705–716

Link AN, Siegel DS, Bozeman B (2017) An empirical analysis of the propensity of academics to engage in formal university technology transfer. In: Universities and the entrepreneurial ecosystem. Edward Elgar, Cheltenham

Meyer JF (2003) Strengthening the Technology capabilities of Louisiana Universities, *Pappas & Associates Report*, Prepared for the Louisiana Department of Economic Development, June 30

Meyer M, Tang P (2007) Exploring the "value" of academic patents: IP management practices in UK universities and their implications for third-stream indicators. Scientometrics 70 (2):415–440

Miller K, McAdam R, McAdam M (2018) A systematic literature review of university technology transfer from a quadruple helix perspective: toward a research agenda. R&D Manag 48(1):7–24

Mowery DC, Nelson RR, Sampat BN, Ziedonis AA (2001) The growth of patenting and licensing by US universities: an assessment of the effects of the Bayh–Dole act of 1980. Research Policy 30(1):99–119

O'Kane C (2018) Technology transfer executives' backwards integration: an examination of interactions between university technology transfer executives and principal investigators. Technovation 76:64–77

O'Kane C, Mangematin V, Geoghegan W, Fitzgerald C (2015) University technology transfer offices: the search for identity to build legitimacy. Res Policy 44(2):421–437

OECD (2001) Innovation and the strategic use of IPR, paper DSTI/STP/TIP 4, Paris

OECD (2003) Turning business into science: patenting and licensing at public research organisations. OECD, Paris

PACEC (2003) *The Cambridge phenomenon—fulfilling the potential*. Report for the greater Cambridge partnership. Public and Corporate Economic Consultants, Cambridge

Sanchez AM, Tejedor ACP (1995) University-industry relationships in peripheral regions: the case of Aragon in Spain. Technovation 15(10):613–625

Schaettgen M, Werp R (1996) Good practice in the transfer of university technology to industry. EC, Brussels

Scott A, Steyn G, Geuna A, Brusoni S, Steinmueller E (2001) The economic returns to basic research and the benefits of university-industry relationships: a literature review and update of findings. Report for the Office of Science and Technology by SPRU—Science and Technology Policy Research

Shattock M (2001) In what way do changing university–industry relations affect the management of higher education institutions. In: Hernes G, Martin M (eds) Management of university-industry linkages. UNESCO/International Institute of Educational Planning, Paris

Sideri K, Panagopoulos A (2018) Setting up a technology commercialization office at a non-entrepreneurial university: an insider's look at practices and culture. J Technol Transf 43 (4):953–965

Siegel DS, Waldman D, Link A (2003) Assessing the impact of organizational practices on the relative productivity of university technology transfer offices: an exploratory study. Res Policy 32(1):27–48

Chapter 2
University Research Commercialisation: Contextual Factors

*In general, the process of commercialising intellectual
property is very complex, highly risky, takes a long time, and
costs much more than you think it will.*
*US Congress, Committee on Science and Technology (1985,
p. 12)*

Abstract University research commercialisation is influenced and driven by macro
and institutional factors that will determine how universities support technology
transfer. Universities in creating and developing technology transfer offices need to
take a proactive strategic approach that is embedded in their local environmental
conditions. The business model framework for TTOs draws together key components that address these contextual factors, as well as creating a strong organisational
posture to support further development and evolution.

Keywords Barriers · Stimulants · Institutional factors · Technology transfer ·
Entrepreneurial · Universities · Industry

2.1 Introduction

University-based research commercialisation is complex and challenging. The reality is that only a very small proportion of the knowledge generated through research
ever reaches a point where it provides a commercial return to the various players in
the progression of an invention through to commercialisation (Forfas 2004: 16).
Consideration of the issues that shape the parameters for commercialisation activities, however, should serve to mitigate some of the risks and help manage expectations with regard to the likely success of commercialisation activities. In this chapter,
we explore some of the macro and micro level stimulants to third stream activities,
before considering the institutional, operational, and cultural barriers that may
constrain technology transfer efforts. Together these factors determine the nature
and extent of technology transfer in universities.

© The Author(s), under exclusive licence to Springer Nature Switzerland AG 2020   15
J. A. Cunningham et al., *Effective Technology Transfer Offices*, SpringerBriefs in
Business, https://doi.org/10.1007/978-3-030-41946-2_2

Table 2.1 Macro level and institutional factors

Macro Level Factors
• Regional R&D
• Government and State Agency Support
• Institutional and Legislative Context
• Features of the Patenting Process
• Legal and IP Systems
Institutional Factors
• Scientific Excellence and Reputation
• Strategy, Mission, and Clear Objectives
• Top Level Leadership and Commitment
• Quality and Effectiveness of Technology Transfer Office
• Age and Experience of TTOs
• Scope of Commercialisation Initiatives
• Documented Policies
• Educational Offerings
• Knowledge of Research and Relations with Faculty
• Networking and Informal Relations
• Trust and Common Expectations
• Inventor Involvement
• Realistic Expectations

2.2   Commercialisation Culture and Ethos: Macro and Institutional Factors

The commercialisation culture and ethos within universities is influenced and shaped by a range of factors. The entrepreneurial university paradigm is accelerating such changes (see Cunningham et al. 2017). Some of these factors TTOs have some control over but many are situational hence they do not have any control. TTOs need to be realistic and pragmatic in creating a commercialisation and ethos that suits these environmental factors. There are macro level and institutional factors that can shape and driver commercialisation culture and ethos (see Table 2.1).

2.3   Motivations and Barriers to Commercialisation Experienced by Researchers

We use Van Dierdonck and Debackere (1988) identified three barriers to commercialisation; institutional barriers (unclear norms and policies, resource, and expertise constraints); operational barriers (constraints on research, motivational issues); and cultural barriers (mutual incomprehension), as a structural framework to discuss the major barriers for commercialisation as depicted by the literature and empirical research in this area.

2.3.1   Institutional Barriers to Technology Transfer

TTO Organisational Structure and Supporting Infrastructure   Typically, university structures have not been designed to carry out technology transfer. Often structures have evolved in an ad hoc manner and are not guided by any specific policy or clear divisions of labour/job descriptions so that the contact point and process of commercialisation is not clear for academics. Historically, TTOs have not been afforded a prominent position in university organisational structure and among the HEIs management team. Further, as a result of historical development work is often overlapped between university administration offices. In this context, it is on rare occasions that offices leverage their knowledge for mutual benefit. Consequently, the process of commercialisation can be impeded by bureaucracy and poor communication between various parties. Often attempts to rectify these problems, by moving TTOs and Research Offices closer, for example are restricted by lack of available property or building constraints as well as lack of expertise and knowledge in this domain.

Resource Constraints and Expertise Deficiencies   In the past, commercialisation activities have not received adequate attention from university top management. As a result, they have traditionally been underfunded and understaffed. Recent developments have brought about changes in this area, but as a result of rapidly evolving technologies and legislative developments, often TTOs do not have sufficient expertise, particularly in-house legal in specific domain arenas. Moreover, something TTO staff can lack the industry domain knowledge to proactively lead the commercialisation process from invention disclosure to patenting and licencing. Thus, very rarely do TTOs adequately fulfil the role of 'science marketer'. This can prove detrimental in the long term as universities can have filed and received patents for which they are receiving no benefits or income as they have not been subsequently licenced to business.

Many TTOs lack sufficient resources, particularly in covering the cost of patent registration and even to maintain their current IP portfolio. Consequently, TTOs have to make commercial and pragmatic decisions in relation to the IP that they will financially support, given costs involved in patenting. The cost of registering an Irish patent is prohibitive for some TTOs. The impact of such expertise and resource deficiencies becomes most obvious when it comes to managing spin-offs and start-ups. The OECD 'Turning Business into Science Report' noted the limited success in this area can often be attributed to the lack of a sound management of technology transfer, as well as insufficient intellectual property and legislative know-how.

Poor Internal Relations   Poor experiences in dealing with the TTO or long time delays in terms of feedback on invention disclosures can reduce the propensity of academics to utilise the resources of the TTO. Scott et al. (2001) note that poor relationships between TTO staff and faculty can be a barrier to commercialisation. Cunningham et al. (2014) found this was a barrier for scientists in the PI role as some reported having less than satisfactory dealing with TTOs.

Lack of Clarity Over Ownership of Intellectual Property Lack of clarity on intellectual property rights can make negotiations longer and more expensive than would otherwise be the case, and in the extreme case prevents them from being completed (Lambert 2003; Sanchez and Tejedor 1995). Barriers are raised by complex proprietary concerns, especially in the case of jointly sponsored research contracts. The approaches of some companies and other funding entities in the intellectual property rights area, such as the pursuit of extensive rights to university background research developed outside the collaboration can prove to be stumbling blocks to successful commercialisation. Moreover, Siegel et al. (2003) found that the private sector expresses frustration with obstacles that impede the process of commercialisation, such as disputes that arise within the university regarding intellectual property rights.

Perceived Conflicts of Interests Perceptions of a conflict of interest can damage the TTO and institution by weakening public trust. Institutional conflicts of interest, also called conflicts of the mission. Some universities invest in start-up firms or accept equity in lieu of royalties on university-held patents, raising concerns that they might become beholden to a company in which they have a financial stake. Thus, TTOs need to be careful about the commercialisation activities that they conduct and their resultant impact in terms of external perceptions of the TTO and the university (RCI 2003).

Overestimating the Value of IP There is a danger that some university leaders might consider technology transfer activities solely as a revenue source, as opposed to a component of the university's mission and overall public and civic responsibility. Such attitudes can raise barriers to negotiations, and actually reduce revenue over the long term. Premature definition and valuation of intellectual property can become an obstacle at the initiation stage of a collaborative project. Attempts at maximisation can contradict the mission of the TTO while at the same time act to the detriment of the commercialisation process (Sanchez and Tejedor 1995).

Partners May Lack Understanding or Trust In some cases partners enter into agreements with an inadequate understanding of the management, internal politics, decision-making structures, and even fundamental interests of the other partners, resulting in slow decisions and insufficient resources. Of the many ingredients in a successful negotiation between companies and universities, mutual trust is perhaps the most important.

Lack of Support for SME Technology Transfer Although most of the firms in a region are likely to be SMEs. Often such firms do not have the personnel, time, or financial resources to invest in technology transfer activities and have very limited knowledge of them. For TTOs, this makes contact and negotiations with SMEs unique and time consuming. However, for those universities that do provide specific SME-related contact points there can be benefits and impacts.

Complicated Technology Transfer Policies One barrier to commercialisation which reduces the propensity of an individual researcher to disclose an invention is the practical difficulties of negotiation and the complex and burdensome nature of the commercialisation process. If policies are not clear or user friendly, staff will be reluctant to follow the commercialisation route, particularly given their time and knowledge constraints. Therefore potential IP will go 'out the back door' and the university does not accrue any of the benefits. Having an effective and efficient invention disclosure process is essential to mitigate against this activity.

2.3.2 Operational Barriers to Research Commercialisation: Constraints and Activities

Lack of Space Pursuing new research activities in most instances requires the recruitment of new researchers, the acquisition of additional physical space, and access to laboratory equipment. With ongoing institutional pressure on physical space and other resources, this means that these resources may not be readily available. This can add another barrier to the successful completion of research goals and this lessens possible research commercialisation opportunities.

Lack of Investment in R&D by Companies A criticism frequently levelled at the industry is that it does not invest sufficient resources in long-term research and development, despite policy initiatives to encourage such investment. The reality is that investment undertaken by companies is largely devoted to overcoming short-term operational issues and focused on cost reduction. This can be a significant barrier to the research commercialisation as it cuts research off at its roots.

Lack of Funding for Prototype Development There also tends to be a funding gap between research and prototype development, which can be a significant barrier to research commercialisation. Researchers can get locked into a vicious circle when the period of a research grant ends, and the associated project is at a stage where it requires resources to take it to the next step to make it commercially attractive for companies. However, in most cases, there maybe no funding available for this activity. Consequently, pressures to attract new research funding force researchers into applying for further research funding, commencing entirely new, possibly unrelated, projects.

Research Treadmill To maintain a reasonable research programme with research staff and postgraduate researchers, senior researchers must constantly submit applications for additional research funding—the research treadmill. A difficulty occurs where there are no funds available for researchers to further develop their existing innovations and they must submit proposals for new basic research when they have not taken previous research to a stage where it is ready for commercialisation or technology transfer. Additionally, the constant requirement to win research funding is a time-consuming process, taking away much of the time available to the researcher that may be used for commercialisation or technology transfer activities.

Constraints for Researchers On an individual level, academic staff can have little time to either establish or undertake collaborative projects with industry in addition to their other duties. Moreover, the continued emphasis on traditional research outputs, such as journal papers, has meant that collaborative industrial R&D is not valued, except as a source of income. Moreover, the general lack of academic recognition for commercialisation and rewards for publications, as opposed to patents, has been a major barrier to the commercialisation research (RCI 2003). As a result, many academics have been faced with the dilemma of either publishing their results for short-term revenue and academic recognition or withholding them until they are patented, with the risk of technology becoming obsolete. Factors limiting the motivational drive of research include:

- *Perceived Publishing Constraints*: Many academics can feel that publication delays and non-disclosure requirements may impair the openness of the university research environment and subsequently affect promotion and tenure decisions. This concern is believed to be particularly strong in the case of younger academics as delays may impair them from building a strong record of publications needed to gain tenure.
- *Age of Academic*: The age of the academic is often an issue in terms of likelihood of commercialisation. Some argue that younger academics are more likely to commercialise as they are less risk averse than their older colleagues. Others suggest that older academics are more likely to undertake commercialisation activities as they have more experience and may not be so 'career orientated'.
- *Contract Researchers and Salaries*: There is an increasing tendency at institutions to employ a high number of research scientists on short- to medium-term contracts. Consequently, due to the lack of permanent staff, these institutions are hindered in their efforts to develop and retain a critical mass of expertise in key research areas. A parallel issue resulting from the high turnover of contract staff is the disappearance from the technology cycle of the individual researcher with the greatest knowledge and understanding of the technology before the technology has reached the commercialisation stage. In environments where there are ceiling on salaries, universities are somewhat constrained in their ability to attract and retain specialist academic staff.

2.3.3 Cultural Barriers: Mutual Incomprehension

Universities and industries are not natural partners; their cultures and their missions are different. Academics value their freedom and independence, resent their reliance

Table 2.2 Cultural differences: universities versus industry

University values	Industry values
New invention	New application
Advancement of knowledge	Added value
New means for further research	Financial returns
Basic research	Applied
Long term	Short term
To know how? What? Why?	Product/service driven
Free public good	Secrecy
Publication	Protection/patents
Academic freedom	Commercial approach
Supply side model of action	Demand side model of action

Obviously, such extreme polarisation gloss over issues that are remarkably diverse and complex such as basic versus applied research

on public funding and feel their efforts are not properly appreciated (Lambert 2003). Reflecting this, the most frequently cited barrier to commercialisation both in the literature and from our research was the cultural barriers that exist between the TTO, the university scientists, and industry (see Cunningham et al. 2014). O'Kane et al. (2017) found that the lack of university support and resources can be a barrier for scientists in the principal investigator role to undertake technology transfer. O'Reilly and Cunningham (2017: 274) found 'personal relationships asset scarcity and proximity issues as barriers and enables to technology transfer engagements with SMEs'. Naturally different stakeholders will have differing expectations, but in the university context such differences are magnified. Research by Sanchez and Tejedor (1995) made reference to managers who rated their relationship with university staff as of little benefit because of the 'the impending impact of a cultural barrier'. Based on our research, we have attempted to capture the main differences between university and industry that creates the mutual incomprehension (see Table 2.2).

While simplifying differences (see Table 2.2), clearly indicates how collaboration between the two parties may prove difficult. Universities, for example are dedicated to academic freedom, publication, and self-determination (Graff et al. 2002). Increasingly universities are adopting an entrepreneurial university paradigm where they need to support entrepreneurship and innovation agendas through the creation of TTOs, dedicated co-operative research centres, etc. (Dolan et al. 2019). Faculty members tend to be concerned with their research agenda and career issues (see Cunningham et al. 2016). This coupled with the pressures that are brought in universities adopting an entrepreneurial paradigm means that it creates further tensions of the individual scientists, particularly in the PI role (Mangematin et al. 2014) and have to deal with associated managerial challenges (Cunningham et al. 2015). Industry, on the other hand, is driven by a clear product-driven focus and a culture that emphasises applied research, secrecy, and protection through patents (Fassin 2000; Nelsen 2001). The focus tends to be on the short-term results and targets, particularly in sectors such as the food sector. Industry can view universities as a low-cost route to solve immediate operational problems but also can benefit by sharing R&D costs with academic partners (see Cunningham and Link 2015). Research by Jones-Evans (1999) comparing ILOs in Sweden and Ireland similarly found cultural differences manifest in different conceptions in terms of time, priorities, and bureaucracy.

Technology transfer will only occur when university faculty and representatives from business and industry work together for mutual gain and find mechanisms to manage the inherent conflict between openness, characteristics of the scientific community and privacy/secrecy, which belongs to the industrial world. Therefore, industry–university collaboration cannot be forced and cultural differences must be understood. More specifically particular commercialisation barriers arise from the following:

Departmental and Faculty Attitudes Towards Commercialisation An attitude can pervade in some university departments and faculties by more established academics that commercialisation is not a worthwhile activity for academics to engage in. This view considers that the research undertaken by individual academics should be available for the public good and is not to be exploited for a substantial return for the academic or their institution. This is a significant barrier to commercialisation. In some instances, this view is held by established academics that hold key positions such as Head of Department or Dean and plays a lead role in the creation of a culture where commercialisation by academics is not valued, encouraged, or supported at departmental, faculty, or institutional levels (Organ and Cunningham 2011). Those academics who proceed to commercialise their intellectual property rights can be regarded as mavericks and may find it difficult to get promoted in the long term. Furthermore, the research training and background of some tenured academic staff means they never experienced academic entrepreneurship in operation. Consequently, their mindsets have not broadened to accept the value of commercialisation undertaken by early career academic staff.

Lack of Awareness Many academics have little knowledge or understanding of the patenting process. The general understanding is that a patent protects one's ideas, however, the patent is very much a lateral application. Therefore, there is an awareness gap in terms of understanding patents and determining their different uses. Furthermore, academics may be unaware they are unable to publish directly related material directly prior to filing for a patent. In addition, there can be a lack of awareness regarding the length of time before patents lapse.

Lack of Soft Skills A deficiency issue in soft skills can also form a barrier to commercialisation activities. These soft skills include communication skills, presentation skills, time management, and the ability to negotiate. These skills particularly come into play when a researcher is undertaking research within industry and also in the adoption phase by industry.

2.4   Measuring Performance

A major barrier to legitimising and gaining buy-in for commercialisation activities stems from difficulties in identifying appropriate activity and outcome indicators. This is a deficiency, although it reflects a general difficulty in that the national systems of innovation that underlie and help to explain performances involve complex phenomena, which are accordingly difficult to measure and analyse. Typically, the most commonly used measures used for TTOs are patents, licences, and spin-outs which are limited and do not fully capture the enormity of organisational effort that goes into commercialising university-based research.

2.4.1   The Difficulty of Measurement

There is a wide range of technology transfer mechanisms available for the commercialisation of research. We acknowledge the complexity and uncertain nature of pursuing any of these technology transfer mechanisms in supporting technology transfer. This complexity necessarily makes attempts to measure the impact of technology transfer activities difficult as there is a potential wide scope of the benefits, many of which are complex and arise through indirect channels, which make causality difficult to attribute (Mowery et al. 2001). Therefore, the true impact of TTO interventions on the technology transfer process is difficult to measure quantitatively (Fassin 2000: 40). This difficulty is accentuated by the heterogeneity in TTO objectives and multiple stakeholder involvement, which can mean success is often the result of complex relationships that involves the whole scientific community (Thursby et al. 2001).

Despite these difficulties, there is a requirement to measure the activities and progress of TTO development and different empirical studies have done this in different country contexts (see Caldera and Debande 2010; Hülsbeck et al. 2013; Vinig and Lips 2015). If these measurements are linked to university-wide objectives they can promote co-operation and a sense of common purpose. Extreme caution, however, must be exercised against measurements that focus exclusively on the revenue-generating activities of technology transfer as:

(a) This will perpetuate the negative image of commercialisation activities among already cynical staff.
(b) This will perpetuate the idea of Technology Transfer as a major source of revenue generation for HEIs. The true revenue generation effects of Technology Transfer, even in the United States, however, are extremely overestimated and in reality quite minimal (RCI 2003).

Royalty income, for example is an important measure of technology transfer performance because of its direct relation to economic impact. Yet in the context of total sponsored research funding, royalty income is a relatively minor contribution to the finances of institutions. In the United States, this is as little as 3% of total institution finances and is thus easily dwarfed by tuition income and charitable donations (Stevens and Phil 2003).

The introduction of indicators cannot be seen to be proposed merely for their role as performance metrics but rather must be promoted to staff for the way in which they will facilitate university administrators in the support and management of activities. Crucially there must be buy-in from TTO staff as such indicators have to be useful to those who will be responsible for their collection, as opposed to being viewed as an external imposition for the sole purpose of measuring and rewarding past performance (Molas-Gallart et al. 2002: 46). In order to develop appropriate indicators there are a number of challenges that need to be recognised and addressed.

2.4.2   Metrics and Indicators: Some Challenges

Need for 'Local' Metrics   There is no one model of the successful university technology transfer. Each university is a product of a distinct process of social, economic, and intellectual development (Guerrero et al. 2014). Developing a set of indicators will require a degree of flexibility in how the indicators and their emphasis are applied to different types of universities. The system of indicators should allow for a 'variety of excellence' to emerge (Molas-Gallart et al. 2002).

Organic Nature of University–Industry Interactions   The relationships between university and industry are often subtle, informal, and linked to personal exchanges between individuals (see Grimpe and Hussinger 2013; Schrader 1991; Link et al. 2017). Past studies have found that many of these interactions are often immune from direct influence by policy or management interventions (Mowery et al. 2001). Universities have traditionally found it difficult to encourage staff to document such interactions and therefore the full extent of university–business interactions, or the benefits stemming from this interaction can prove difficult to measure. This organic nature of interaction limits the ability to find suitable measures and instruments to shape behaviour. It is possible that there will be no one-to-one matching between policy actions and interaction patterns (Molas-Gallart et al. 2002). Research by Sanchez and Tejedor (1995) has indicated that links are often established informally without the assistance of a TTO.

Differences Across Disciplines   The level, extent, and nature of interaction and commercialisation mechanisms will differ by discipline. There will naturally be differences between applied disciplines (like mechanical engineering, business administration, or medicine) and fundamental theoretical disciplines (like theoretical physics or philosophy). While in the former, direct channels of application may exist, in the latter the impact of academic activities on the economy and social welfare is likely to be more long term and indirect. Measurement of technology transfer will need to be all encompassing in order to capture all these activities and effects.

Resistance/Lack of Enthusiasm for Evaluative Metrics There is little enthusiasm in universities for new measurement systems. Often such systems can be seen as a resource constraint and administrative burden, or staff feel as if such metrics should be reserved for private sector organisations. Promotion of the benefits and objectives in setting up such metrics is therefore critical particularly around planning supports that will be of actual benefit to the academic community.

Interdependence Between University Activities Technology transfer success is dependent on other activities within universities, particularly research. Also, effective organisational structures to support technology transfer can play a role in determining technology transfer performance. In viewing and designing technology transfer performance, universities need to understand and appreciate the interdependent with other activities as well as institutional strengths.

Indirect Impacts Related to the above technology transfer activities may also result indirectly in research contracts, grants, and donations to the university, as well as an enhanced student intake as a result of reputation/linkages with companies (Graff et al. 2002).

Serendipity The outcome, and therefore the impact, of research activities are by their very nature unpredictable, and serendipity is an important element when companies are attempting to develop new products and generate competitive advantage.

Efficiency *versus* **Effectiveness** It is often the case that measurement systems focus on quantitative output and measurement can be reduced to looking at numbers, e.g. invention disclosures, patents applied for, patents granted. While useful such metrics do not capture the effectiveness of the process, e.g. the extent and nature of interaction between TTO staff and academics, or the quality of the activities conducted, e.g. evaluating an invention disclosure in terms of its patentability.

2.5 Business Model Framework for Technology Transfer Offices

Taking the business model (Baden-Fuller and Morgan 2010; Casadesus-Masanell and Ricart 2010; Magretta 2002; Chesbrough 2007; Teece 2010) perspective we develop a business model framework for TTOs. International best practice evidence provides a basis for understanding the requirements for success, individual universities will have to find their own solutions to these issues, depending on their size, disciplinary base, the economic and social strengths, and needs of their region (Hall 2004; Shattock 2001). Moreover, Baglieri et al. (2018) in their study of 60 US universities found four distinctive technology transfer business models. This study affirms differences and distinctiveness among TTOs. The basis of such developments should, therefore, acknowledge the 'variety of excellence' that exists and

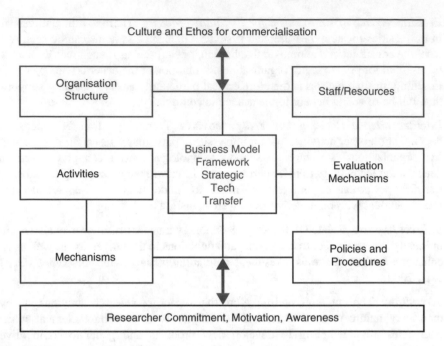

Fig. 2.1 Business model framework for technology transfer offices

allow each university to choose its own distinctive strategy (Molas-Gallart et al. 2002). As the OECD (2003: 13) notes: 'there is no one size fits all approach to technology transfer.' In recognising this, our business model framework for technology transfer offices draws on elements of best practice elsewhere, acknowledging that policy choice and that implementation needs to be adapted to suit local circumstances and contingencies. Key components of your business model framework for technology transfer offices are presented in Fig. 2.1.

The central issue is that universities must take a more proactive strategic approach to commercialisation and technology transfer activities rather than merely playing a facilitator/information broker role. Surrounding circles indicate the interrelated mechanisms that will support and facilitate this strategic approach.[1] Crucially, this business model framework acknowledges that such changes cannot take place without due consideration to the motivation and awareness of researchers and a supportive ethos for commercialisation in terms of the broader context of the university. Ultimately improving technology transfer and commercialisation are

---

[1]It must be acknowledged that some of the issues suggested will be connected to the types of pressures that will emanate from Government agencies and funding bodies. The authors however recommend a pro-active approach to implementing such policies rather than having them imposed. This relates to rationale that successful transfer stems from experience and older policies. Hence the earlier policies are put in place the more benefit in terms of long-term outcomes.

keys to an innovative economy leading to wealth generation and job creation (Large et al. 2000).

An Issue of Balance

Crucially while commercialisation of intellectual property owned by universities is an important component of third stream activities and of entrepreneurial universities, it is only one among the many other functions that link universities with society (Molas-Gallart et al. 2002). In this respect, it is worth remembering the following:

- The naturally occurring, organic nature of university–industry interactions.
- The fact that third stream and second stream (research) activities are not independent.
- The considerable difference between universities across disciplines in the ways that research influences society.

Appreciation and acknowledgement of these issues can facilitate in preventing any potential backlash by researchers or staff who may feel commercialisation activities are being imposed on them (OECD 2003). Essentially the issue is one of balance. The emphasis should be to balance the objectives of managing intellectual property rights, developing new revenue sources, and accommodating faculty interests while simultaneously maintaining norms related to the conduct of academic research and the dissemination of research findings (Feldman et al. 2002; OECD 2003: 104).

2.6   Some Implications for Universities and TTOs

The key issues highlighted in this chapter, which need to be considered by HEIs include:

Stimulants for technology transfer and research commercialisation, which include **macro level factors** such as:

Macro Level Factors

- Regional R&D
- Government & State Agency Support
- Institutional and Legislative Context
- Features of the Patenting Process
- Legal and IP Systems

Institutional Factors

- Scientific Excellence and Reputation
- Strategy, Mission, and Clear Objectives
- Top Level Leadership and Commitment
- Quality and Effectiveness of Technology Transfer Office
- Age and Experience of TTOs

- Scope of Commercialisation Initiatives
- Documented Policies
- Educational Offerings
- Knowledge of Research and Relations with Faculty
- Networking and Informal Relations
- Trust and Common Expectations
- Inventor Involvement
- Realistic Expectations

Three types of barriers to commercialisation of research are institutional barriers, operational barriers, and cultural barriers.

Institutional Barriers

- TTO organisational structure and supporting infrastructure
- Resource constraints and expertise deficiencies
- Poor internal relations
- Lack of clarity over ownership of IP
- Perceived conflicts of interests
- Overestimating the value of IP
- Patenting of research tools may discourage beneficial research
- Partners may lack understanding or trust
- Lack of support for SME technology transfer
- Complicated technology transfer policies

Operational Barriers to Research Commercialisation

- Lack of space
- Lack of investment in R&D by companies
- Lack of funding for prototype development
- 'Research treadmill'
- Availability of researchers to pursue research fellowship
- Constraints for researchers
- Contract research and salaries
- Lack of full-time research positions

Cultural Barriers

- Department and faculty attitudes towards commercialisation
- Lack of awareness
- Perceived non-challenging nature of commercialisation
- Lack of soft skills

Evaluating Technology Transfer
The true impact of technology transfer is difficult to measure quantitatively (Fassin 2000: 40). Caution must be exercised against measurements that focus on revenue generating activities of TTOs.

- Introduction of indicators cannot be seen to be proposed merely for their role as performance metric but rather must be promoted to staff for the way in which the indicator will facilitate university administrators in the support and management of activities. Moreover, metrics should attempt to represent the full extent and a wide range of knowledge transfer that is actually transferred.
  Consideration should be given to the following factors when developing metrics to measure technology transfer activities:

  - Need for 'local' metrics
  - Organic nature of university–industry interactions
  - Differences across disciplines
  - Resistance/lack of enthusiasm for evaluative metrics
  - Interdependence between university activities
  - Indirect impacts
  - Serendipity
  - Overemphasis on efficiency versus effectiveness

- Quantitative measures for TTOs could include:

  - Research expenditure
  - Licence option agreements
  - Research funding
  - Licence income
  - Legal fees, expenditures, and reimbursements
  - Patent-related activity
  - Start-up companies
  - Licenced technologies and post-licencing activity

- There needs to be a complementary set of evaluation mechanisms that evaluate technology transfer activity *and* impact.

References

Baden-Fuller C, Morgan MS (2010) Business models as models. Long Range Plan 43 (2–3):156–171

Baglieri D, Baldi F, Tucci CL (2018) University technology transfer office business models: one size does not fit all. Technovation 76:51–63

Caldera A, Debande O (2010) Performance of Spanish universities in technology transfer: an empirical analysis. Res Policy 39(9):1160–1173

Casadesus-Masanell R, Ricart JE (2010) From strategy to business models and onto tactics. Long Range Plan 43(2–3):195–215

Chesbrough HW (2007) Why companies should have open business models. MIT Sloan Manag Rev 48(2):22

Cunningham JA, Link AN (2015) Fostering university-industry R&D collaborations in European Union countries. Int Entrep Manag J 11(4):849–860

Cunningham J, O'Reilly P, O'Kane C, Mangematin V (2014) The inhibiting factors that principal investigators experience in leading publicly funded research. J Technol Transf 39(1):93–110

Cunningham J, O'Kane C, O'Reilly P, Mangematin V (2015) Managerial challenges of publicly funded principal investigators. Int J Technol Manag 68(3–4):176–201

Cunningham JA, Mangematin V, O'Kane C, O'Reilly P (2016) At the frontiers of scientific advancement: the factors that influence scientists to become or choose to become publicly funded principal investigators. J Technol Transf 41(4):778–797

Cunningham JA, Guerrero M, Urbano D (2017) Entrepreneurial universities—overview, reflections, and future research agendas. In: Siegel D (ed) The world scientific reference on entrepreneurship. Volume 1: Entrepreneurial universities technology and knowledge transfer. World Scientific Publishing, Hackensack, pp 3–19

Dolan B, Cunningham JA, Menter M, McGregor C (2019) The role and function of cooperative research centers in entrepreneurial universities: a micro level perspective. Manag Decis 57:3406

Fassin Y (2000) The strategic role of university-industry liaison offices. J Res Adm 1(2):31–41

Forfas (2004) From research to the marketplace: patent registration and technology transfer in Ireland, Forfas

Feldman M, Feller I, Bercovitz J, Burton R (2002) Equity and the technology transfer strategies of American research universities. Manag Sci 48(1):105–121

Graff G, Heiman A, Zilberman D, Castillo F, Parker D (2002) Universities, technology transfer and industrial R&D. In: Evenson RE, Santianello V, Zilberman D (eds) Economic and social issues in agricultural biotechnology. CABI Publishing, New York, pp 93–117

Grimpe C, Hussinger K (2013) Formal and informal knowledge and technology transfer from academia to industry: complementarity effects and innovation performance. Ind Innov 20 (8):683–700

Guerrero M, Urbano D, Cunningham J, Organ D (2014) Entrepreneurial universities in two European regions: a case study comparison. J Technol Transf 39(3):415–434

Hall, B. H. (2004). University-industry research partnerships in the United States. https://cadmus. eui.eu/bitstream/handle/1814/1897/ECO2004-14.pdf

Hülsbeck M, Lehmann EE, Starnecker A (2013) Performance of technology transfer offices in Germany. J Technol Transf 38(3):199–215

Jones-Evans D, Klofsten M, Andersson E, Pandya D (1999) Creating a bridge between university and industry in small European countries: the role of the industrial liaison office. R&D Manag 29(1):47–56

Lambert R (2003) Lambert review of business-university collaboration. https://papers.ssrn.com/ sol3/papers.cfm?abstract_id=1509981

Large D, Belinko K, Kalligatsi K (2000) Building successful technology commercialization teams: pilot empirical support for the theory of cascading commitment. J Technol Transf 25 (2):169–180

Link AN, Siegel DS, Bozeman B (2017) An empirical analysis of the propensity of academics to engage in formal university technology transfer. In: Universities and the entrepreneurial ecosystem. Edward Elgar, Cheltenham

Magretta J (2002) Why business models matter. Harvard Business Review

Mangematin V, O'Reilly P, Cunningham J (2014) PIs as boundary spanners, science and market shapers. J Technol Transf 39(1):1–10

Molas-Gallart J, Salter A, Patel P, Scott A, Duran X (2002) Measuring third stream activities. Final report to the Russell Group of Universities. SPRU, University of Sussex, Brighton

Mowery DC, Nelson RR, Sampat BN, Ziedonis AA (2001) The growth of patenting and licensing by US universities: an assessment of the effects of the Bayh–Dole act of 1980. Res Policy 30 (1):99–119

Nelsen L (2001) University technology transfer practices: reconciling the academic and commercial interests in data access and use. http://www.nap.edu/html/codata_2nd/ch13.html

O'Kane C, Zhang JA, Cunningham JA, O'Reilly P (2017) What factors inhibit publicly funded principal investigators' commercialization activities? Small Enterp Res 24(3):215–232

O'Reilly P, Cunningham JA (2017) Enablers and barriers to university technology transfer engagements with small-and medium-sized enterprises: perspectives of principal investigators. Small Enterp Res 24(3):274–289

OECD (2003) Turning business into science: patenting and licensing at public research organisations. OECD, Paris

Organ DJ, Cunningham J (2011) Entrepreneurship in the academy: the case for a micro-institutional analysis. DRUID 2011 on innovation, strategy, and structure—organizations, institutions, systems and regions at Copenhagen Business School, Denmark, 15–17 June 2011

Sanchez AM, Tejedor ACP (1995) University-industry relationships in peripheral regions: the case of Aragon in Spain. Technovation 15(10):613–625

Schrader S (1991) Informal technology transfer between firms: cooperation through information trading. Res Policy 20(2):153–170

Scott A, Steyn G, Geuna A, Brusoni S, Steinmueller E (2001) The economic returns to basic research and the benefits of university-industry relationships: a literature review and update of findings. http://sro.sussex.ac.uk/id/eprint/18177/

Shattock M (2001) In what way do changing university–industry relations affect the management of higher education institutions. In Part III, in Hernes G, Martin M (eds), Management of University-Industry Linkages, Policy Forum (No. 11)

Siegel DS, Waldman D, Link A (2003) Assessing the impact of organizational practices on the relative productivity of university technology transfer offices: an exploratory study. Res Policy 32(1):27–48

Stevens A, Phil D (2003) 20 years of academic licensing—royalty income and economic impact. J Licensing Exec Soc Int (les Nouvelles) 38:133–140

Teece DJ (2010) Business models, business strategy and innovation. Long Range Plan 43 (2–3):172–194

Thursby JG, Jensen R, Thursby MC (2001) Objectives, characteristics and outcomes of university licensing: a survey of major US universities. J Technol Transf 26(1–2):59–72

University-Industry Collaboration Initiative (RCI) (2003) Working together, creating knowledge. Business Higher Education Forum

US Congress (1985) Hearings of the committe on science & technology, US house of representatives, 98th Congress, second session, 21 March 1984

Van Dierdonck R, Debackere K (1988) Academic entrepreneurship at Belgian universities. R&D Manag 18(4):341–353

Vinig T, Lips D (2015) Measuring the performance of university technology transfer using meta data approach: the case of Dutch universities. J Technol Transf 40(6):1034–1049

Chapter 3
Business Model Framework: Strategic Considerations

Abstract Business model framework strategic considerations focus on the development of a strategic plan, mission statement, and strategic priorities. TTOs focus more on action as opposed to facilitation and developing an organisational structure that is appropriate for the institutional context, as there is no one best way to structure TTO activities. The chosen organisational structure of TTOs is heavily shaped by institutional factors and their strategic focus. Strong leadership, investment in the skills and capabilities of technology transfer professionals, and configuring a sustainable resource base to support services are critical factors for TTO effectiveness. TTOs also need to consider how best to cooperate with other TTOs to potentially reduce any potential resource and expert deficiencies and leveraging economies of scale.

Keywords Strategic plans · Organisation structure · Skills · Resources · Expertise · Technology transfer

3.1 Introduction

TTOs have become more commonplace in university institutional settings. If managed well and resourced appropriately, they can be a significant institutional asset and enable researchers to realise the commercial potential of their research endeavours. The first core element of our business model framework is strategic consideration, which sets the direction and scope of activities of TTOs. We firstly focus on strategy and in particular, strategic plans and mission statements that are the artefacts that are used to mobilise support and ultimately resources from stakeholders within and outside the university. We then explore technology transfer activities in terms of platforms, developing active networks, core competencies, and boundary-spanning activities. TTOs also need to configure an organisational structure that best fits their institutional environment, scope, and activities. We focus on the oversight structures, location of offices, autonomy, specialism, and different strategies that TTO can adopt to overcome resource constraints such as shared services. Our final focus in this chapter is on expertise and resources particularly with respect to strong

© The Author(s), under exclusive licence to Springer Nature Switzerland AG 2020   33
J. A. Cunningham et al., *Effective Technology Transfer Offices*, SpringerBriefs in
Business, https://doi.org/10.1007/978-3-030-41946-2_3

leadership, skills, HR strategy integration, and resources. These are strategic considerations that TTO needs to focus on as they evolve and grow.

3.2   Strategic Plan

The emergence and continual development of TTOs at some universities is very much in a passive or reactive manner. The downfall of such a reactive rather than strategic and proactive approach tends to be lower levels of patenting and licencing, coupled with poor infrastructural support and low levels of awareness and commitment to commercialisation activities (OECD 2003: 42). Universities should, therefore, develop a more proactive, strategic approach, to technology transfer, including the development of a mission statement, a strategic plan, and the identification of strategic priorities.

> **Key Success Factor**
> Development of a Strategic Plan, Mission Statement, and Strategic Priorities.

3.2.1   Developing Strategic Plans

As universities become more involved in commercial activities of one kind or another, they will have to develop a clearer sense of their mission, their objectives, and develop clear policies to facilitate implementation. As a result of the universities' far-reaching role, activities are often uncoordinated and subject to burdensome systems of accountability and regulation. The TTO should have a distinctive strategic plan that is aligned with the overall university mission and strategic priorities. Such a plan should define the goals and mission for technology transfer and in so doing provide a sense of direction and leadership as well as a clear rationale for action.

Universities must provide the organisational support to attract, negotiate, and carry out collaborations with industry. This needs to be assessed and documented in order for various parts of administrative structures to work together in an effective manner to reach common goals. Critical would be to clearly link objectives into the university's strategic and academic plans, and demonstrate how activities will facilitate in achieving aspects of the university's own mission.

While TTOs are more complex to manage than businesses they need to do business like in the way they manage their affairs (Lambert 2003). Aragonés-Beltrán et al. (2017) note: 'This growth has meant in many cases the incorporation of the TTO activity into the strategic planning process of the university and therefore the definition of clear objectives in technology and knowledge transfer activities as well as the resources and actions devoted to carry out the activities designed to fulfil those objectives.' Strategic management of relations implies, not only the formalisation of policy priorities, but also top-level management support (Hernes and Martin 2001). The commitment of the institution's leadership is essential to the long-term success of technology transfer initiatives. A strategic plan serves to galvanise, and provides evidence of leadership and commitment. Such evidence can be particularly useful in contexts where university technology transfer is controversial. Once a plan has been devised, the institution's leadership will also play a key role in communicating the objectives and emphasising the importance and relevance importance of industry collaborations and technology transfer (Meyer 2003).[1] Aside from communication on websites and speeches as well as celebrating success stories, successful university change programmes have involved an annual letter to the research faculty and staff, written by the Chancellor and President highlighting the importance and appropriateness of involvement in technology transfer (Meyer 2003). Such tangible and visible commitments from the university top management team along with a resource commitment can contribute to changes and embracement of research commercialisation among faculty (see Chang et al. 2009).

Crucially the process of developing the key objectives and mission statements must be an all-inclusive one. Acceptance and agreement of objectives by multiple stakeholders is of crucial importance, and also serves to facilitate the process of implementation. Reviewing best practice in technology transfer, Allan (2001) noted that development or reform of policies should involve inputs from funding agencies, scientists, and industrial partners as well as other actors in the technology transfer chain. Stakeholder involvement will not only create buy-in to the newly established objectives but also ensure that there is a consensus in terms of expectations of the costs, risks, and time lag associated with commercialisation activities.

Objectives arising from the planning process should be clear and, while they must take historical perspectives and political realities that bear upon the particular institution into consideration, policies and objectives should be as simple and straightforward as the institution's circumstances permit. Understandable objectives in the form of a strategic document can ensure effective management and provide the TTO with legitimacy when dealing with internal and external stakeholders. A small number of important objectives (no more than 7–10) also make communication and dissemination of policy easier (see Table 3.1).

3.2.2   Developing Mission Statements

The development of a strategic plan, however, can only take place in an effort to meet the aims and scope of activities as laid out as part of a mission statement. A

---

[1]Allan (2001) and Scott et al. (2001) note that one mechanisms for such emphasis is that the language of technology transfer is an inherent part of public statements, publications, and speeches made by, and on behalf of, the leadership of an institution.

Table 3.1 Elements of the strategic plan to support technology transfer for Penn State University

Goal 6: Invigorating entrepreneurship, technology transfer, and economic development
SIGNATURE OBJECTIVE 6.1 Establish Penn State as the "go-to" academic institution for industry-sponsored research by creating a local ecosystem where Penn State researchers work hand-in-hand with industry to accelerate technology transfer
OBJECTIVE 6.2 Create a Penn State culture that values entrepreneurship, technology transfer, and economic development
OBJECTIVE 6.3 Improve, streamline and clarify processes to accelerate the pace at which technologies move from discovery to implementation

Source: Penn State University (2016) OVPR Strategic Plan, pp. 76–81 (see https://www.research. psu.edu/sites/default/files/OVPR%202015%20Final%20Strategic%20Plan_January_2016.pdf)

mission statement quite simply should communicate what the TTO is and what it does. The statement should avoid any direct reference to the commercialisation of research for fundraising or revenue maximisation purposes as the key role of the TTO. The statement should serve to ease the fears of those who are critical of commercialisation emphasising that technology transfer is a contribution towards fulfilling the overall mission of the university. Of the mission statements reviewed in Chap. 1 most stressed the exploitation of research for economic and/or public good. For example, Penn State University the Office of Technology Management states: 'Our mission is to protect Penn State intellectual property, identify its commercial potential, and stimulate economic development through the transfer of Penn State technologies to the marketplace. The Office of Technology Management also promotes Penn State technology by protecting, marketing, and licensing University inventions to companies for further development and commercialization'.

Such a mission statement neatly captures the dual responsibility of TTOs. Firstly as guardians of university IP, and secondly its role in disseminating this to industry and beyond. Research by Siegel et al. (2003a, b) concluded that mission statements were indicative of leadership and an entrepreneurial climate which was complementary to an HEI generating more licences. Overall, a strategic focus on technology transfer should become manifest in the other interrelated mechanisms that will support and facilitate this approach, including the identification of strategic priorities as elaborated upon below.

3.3   Technology Transfer Activities

Resource constraints acknowledged, successful TTOs have been noted for conducting strategic prioritisation in order to encourage and facilitate research. Allan's (2001) review of best practice suggests that TTOs should try and shed responsibility for everything except the key processes involved in technology transfer. This has been referred to as focusing on core competencies, i.e. activities or areas where the institution excels (or has the ability to excel in).

Key Success Factor
Focused more on Action as opposed to Facilitation.

3.3.1 Identifying Technology Platforms

While the desire to manage all research activities with the same level of effort and commitment will always be to the fore the reality is that most institutions identify areas of competitive strength in research. The salience of research in university portfolios has led to the strengthening of strategic research planning within institutions. These efforts have been guided, in no small way, by the level and type of funding being allocated through government agencies. An example is the Irish university system where the establishment of the Higher Education Authority's Programme for Research in Third Level Institutions and the creation of Science Foundation Ireland has shaped HEIs to develop research bases, promoted research excellence, and encouraged more high-risk fundamental research (see Cunningham and Golden 2015; Geoghegan and Pontikakis 2008; Geoghegan 2017). It takes time for institutions to build critical mass or expertise, or even resources to conduct in-depth research in all areas of science and technology. There is a need for focus and for expenditure to be prioritised within an overall coherent framework that promotes national development objectives (Lambert 2003). In this way universities can build up clusters of knowledge in particular areas, which lend themselves well to commercialisation, should researchers wish to follow this route.

TTO's Role
The TTO should have a central role in identifying and encouraging the development of such strategic research competencies, or what have been termed 'technology platforms'. The TTO should also be in a position to communicate elements of this focus and in turn direct industry making contact with the university to these niches of expertise. Such technology platforms also provide useful opportunities for collaborative research contracts, so that the funding and development of such technology platforms do not remain the sole responsibility of the university. It is often the case that technologies developed in such platforms are clearly associated with the research requirements of an enterprise. This serves to enhance the potential for technology spillovers and increased networking between the industry and the university. An example of this is NUI Galway in the West of Ireland has built an international research reputation in medical devices that supports a significant international industrial cluster (see Cunningham et al. 2015; Green et al. 2001; Giblin 2011; Giblin and Ryan 2015). This has led to the establishment of CURAM a Science Foundation Ireland Centre for Research Medical Devices and a further deepening of university–industry collaboration.

Monitoring Research Activity
TTOs can be proactive and be involved or at least be aware of the research activities
and outputs of the university research community. While traditionally the academic
has taken the lead in approaching the TTO for guidance, best practice suggests that
the TTO should actively monitor the research being conducted in an institution.
Although technically a step removed from the actual research, the TTO should have
familiarity and good contacts with the faculty-based expertise and track potential
industrial research opportunities. Mechanisms that may facilitate this monitoring
include the creation/maintenance of databases of active research in the university as
well as areas of academic specialisms. This mechanism would be useful when it
comes to negotiating and setting up collaborative research contracts or materials
transfer agreements. Furthermore, such approaches maybe useful internally in terms
of linking researchers who may be looking at particular areas of common interest or
with related benefits. Active monitoring is particularly warranted as universities
extend the nature and extent of their research activities.

3.3.2 Developing and Sustaining an Active Network

A central mechanism in providing the opportunity for and facilitating technology
transfer includes the development and sustaining of informal and formal contacts
and networks. TTOs should leverage alumni networks to build closer relationships
with graduates working in the business community. The extent of informality of
these contacts should not be judged as indicative of less effective contact mechanisms, but rather as the OECD (2003) highlighted, informal contacts often form the
initial underpinning to more advanced agreements.

Develop a Presence in the Region
To maximise the potential of this occurrence TTOs should allocate time for staff to
attend the conferences, presentations in their region, nationally and internationally,
and also sustain and build networks with different quadruple helix stakeholders.
Such activities can support horizon scanning in different domain areas and provide
insights that support researchers seeking to commercialise their research.

Horizontal Networking
Networking should not just occur with commercial entities but also horizontally
between universities. Communication and discussion with other universities, and
other TTOs in particular, should facilitate diffusion of best practice as well as
ensuring that staff are up to date with developments in various fields of research.
This horizontal collaboration may also allow smaller universities to learn from larger
best practice institutions. Extensive networking may also lead to the sharing of
knowledge (e.g. knowledge of the patenting process), or the sharing of resources
in order to deal with common problems (e.g. difficulties in interpreting legislation or
facilitate in rectifying expertise deficiencies). Such a body of integrated knowledge
and expertise may also prove more attractive to businesses, particularly

multinationals. The frequency and extent of such interaction maybe formalised if there is a sufficient critical mass and willingness to meet on a more structured and regular basis. Indeed, strategies of alliances and networking have become a key factor behind the success of universities (Lundvall 2002). There should, therefore, be a maximum opportunity for networking, both formally and informally, between TTOs (particularly their specialists in enterprise development and technology licencing). Knowledge Transfer Ireland (KTI) is an example of how at national level knowledge and expertise sharing can be collectively diffused for the benefit of individual TTOs and well as the companies they engage with for technology transfer. KTI has developed a catalogue of model agreements that can form the basis for negotiations between universities and industry.[2] Moreover, there are different professional bodies that university-based TTO personnel such as the Association of University Technology Managers (ATUM), Praxis, ASTP, and Licencing Executives Society (LES). Overall, TTOs need to be more strategic in their approach, moving from random individual contacts and one person focus points to making such interaction an inherent part of their activities and policies.

3.3.3 Focus on Core Activities and Developing Core Competencies

Research from best practice institutions stresses that as TTOs develop they need to become more strategic in their focus and develop expertise in specific areas in which they wish to excel (Brint 2005). In order to reduce the likelihood of being overburdened and missing commercialisation opportunities, there is a necessity to shed non-core activities so that resources and attention can be focused on previously identified strategic priorities. Fassin (2000) emphasises the need for TTOs to focus on their core competency, i.e. *making deals*. Where possible TTOs should minimise academic obligations and shed responsibilities for everything except technology transfer (MTAs, contracts, etc.). In the process of developing a strategic plan, activities should be prioritised in the context of resource constraints and the previously identified technology platforms and research priorities.

Research by Jones-Evans et al. (1999) has highlighted the more sophisticated approach of TTOs in the Swedish context. A longitudinal study of Chinese universities found that they create value through 'dynamic management and active orchestration of assets' (Yuan et al. 2018). While TTOs still serve as one stop shop, in practice this means that certain requests are referred to other offices/centres within the university, which have taken over and developed expertise in these areas such as career placement, student placement, etc. Universities and research institutions cannot be everything and must differentiate themselves and use their TTO to exploit

---

[2]See   https://www.knowledgetransferireland.com/Model-Agreements/Catalogue-of-Model-Agreements/

the research that may prove most valuable. This means that they can focus on developing strong networks and build alliance capabilities that support effective technology transfer (Leischnig and Geigenmüller 2020). As Muscio (2010: 199) notes: 'Managing a TTO requires special skills to facilitate the matching of academic knowledge, competencies and resources to business needs, and provide assistance in the commercialization and pricing of technology. The involvement of professional, non-academic managers in TTOs will support these activities and help to bridge the cultural gap between university and industry'. Moreover, with experienced staff and well-established policies are practices are more likely to attain better technology transfer results (González-Pernía et al. 2013). Therefore, TTOs need to consistently invest in developing individual and collective competencies and skills as well as putting in place clear policies and procedures that support TTOs.

3.3.4 Boundary Spanning: Communicating to Stakeholders and Managing Expectations

Best practice research of TTOs by Allan (2001) suggests that boundary-spanning activities by TTO staff are crucial in maintaining positive relationships with academics and can serve to reduce informational and cultural barriers to research commercialisation. Boundary spanning refers to the specific type of marketing relevant to TTOs.[3] Boundary-spanning capabilities and awareness ensure that TTOs act as an effective bridge between commercial entities and researchers. Research by Siegel et al. (2003a, b) found that boundary spanning on the part of the TTO is crucial and involves adept communication with stakeholder groups in order to forge alliances between researchers and firms.

Communication capabilities are an inherent part of boundary-spanning capabilities. Clear communication is critical for TTOs given their interface role and requirements to serve multiple stakeholder demands and requirements. Communication from the boundary-spanning activities reduces the probability of conflict between different stakeholders as they facilitate in forming clear expectations and an understanding of each other's role (Beesley 2003). Boundary-spanning activities, therefore, prove an effective tool for managing communication and consequently can be crucial in obtaining buy-in at all institutional levels. Furthermore, effective boundary spanning and relationship management enables trust to be built up between partners.

Development of Databases
One mechanism to facilitate and focus boundary-spanning activities is the use of software or a database to track and manage inventions, patents, agreements, and

---

[3]The notion of boundary spanning captures the importance of the networks and contacts of TTOs as in this context advertising or technology brokers are rarely used.

Table 3.2   TTO outreach and awareness raising activities

• Encouraging scientists to develop business skills through enterprise workshops and courses
• Providing role models of success stories, and generate positive messages about the entrepreneurial culture
• Stimulating interaction between industry and research through sabbaticals in the industry for researchers, and by bringing industry people into universities as guest lecturers, members of school boards, etc.
• Building awareness of what is needed to translate a research outcome into the marketplace
• Putting in place appropriate industry-focused boards for R&D groups

contracts. There are a variety of software products in the market place that support this type of activity. Research by Sanchez and Tejedor (1995) has indicated the crucial developmental role of a database network in that it allows staff to know about technology services and research activities available from every university department. Such moves towards a planned organisation can reduce the chance element involved in meeting with contacts and commercialising research.

Outreach and Awareness Raising

One of the key boundary-spanning activities TTOs should conduct is outreach and awareness raising. Such activities could include facilitating periodic meetings, educational seminars, and training sessions for inventors and staff (see Table 3.2).

The goal of such activities is to develop attitudes and understanding of commercialisation. These efforts may be furthered by 'town hall' meetings/sessions involving all stakeholders. As noted by Meyer (2003: 14): 'to ensure success of the technology transfer office, its personnel should make outreach presentations on an ongoing basis and provide educational materials to teach the process of technology transfer to the faculty and student researchers'. Research has shown that such educational dissemination is best made to small groups in an informal interactive setting such as a departmental or college faculty meeting. Presentation content can include the following: the technology transfer process, best practice, TTO role and services, technology transfer mechanisms, and institutional examples of successful research commercialisation. TTOs should use both formal and informal channels and approaches to convey what they do and their success stories. Göktepe-Hultén (2010: 50–51) suggests: 'TTOs should also publish articles about its program and successful stories in the university's in-house publication. These publications may include information on department's technology transfer performance. TTOs can be part of faculty in distribution list for periodic e-newsletter. They can find post-docs or students interested in technology transfer and train them so they can become a liaison between the TTO office and the faculty'.

TTOs must, therefore, spend a portion of their time engaging in boundary-spanning activities to encourage faculty members to disclose inventions. Often the activities and purpose, of staff of the TTO are not known within the university community. Contact and educational initiatives facilitate in creating an image for TTO that stimulates awareness of the services provided and may create interest among researchers. Furthermore, technology transfer is a difficult and lengthy process, requiring patience and persistence. Education and awareness raising can also serve as mechanisms to convey these truths and dispel myths to researchers who possess unrealistic expectations. Such boundary spanning might be getting involved in educational programmes that help develop commercialisation skills and awareness (see Mosey et al. 2006). The impact of such activities can pay back over the medium to long-term. Ultimately the best marketing and communication tool is a good success story. Similarly, the best internal marketing for the TTO is a happy end user of their services. (Fassin 2000: 36). Overall such boundary-spanning activities are essential if the TTO is to prevent researchers bypassing them as found Goel and Göktepe-Hultén (2018: 254): 'Results, based on a large sample of German researchers, show that academic inventors' professional experience, doctoral degrees and scientific discipline empower them to choose to not use TTOs, as do their various industry interactions'. TTOs need to be proactive in their boundary-spanning activities particularly that are internally institutionally focused.

3.4   Organisational Structure

As a reflection of the ad hoc and rapid nature of the development of TTOs, very little attention has been directed at the organisational structure that may best facilitate technology transfer. Many existing offices can be seen to have grown into their current structure rather than have taken a strategic stance to dictate the nature of the structure. Structural options, reporting levels, and levels of autonomy will be heavily shaped by institutional norms and path dependencies. The right mechanisms may depend upon the scale and substantive focus of the collaboration.

Key Success Factors
There is no one best way to structure TTO activities.
   The structure will be heavily shaped by institutional factors and strategic focus.

The structure of TTOs at HEIs are complex and path dependent (Bercovitz et al. 2001; OECD 2003). Historical appreciation of this is crucial; particularly in estimating potential cultural resistance should structural alterations be attempted. The appropriateness of one institutional arrangement or another depends on the context within which the PRO operates: 'its status as a private or public institution, the amount of government funding it receives; the size of its research portfolio and fields of specialisation; its geographical proximity to forms and insertion in innovation networks; and its funding capacity' (OECD 2002).
   As a result of the increased attention that has been directed at technology transfer the requirement for TTOs to have a clear structure and reporting relationships has come to the fore. Shattock (2001) argues that because TTOs have become central to HEI   success,   and   as   a   consequence   of   the   extent   of   university–industry

Table 3.3 Guidelines for developing an appropriate TTO organisational structure

• Ensure clear lines of reporting and responsibility
• Ensure close interaction between commercialisation and research activity
• Typically the TTO director reports to a Vice President, Vice Provost, or Vice Chancellor for Research (Siegel et al. 2003a, b). Improve communication and teamwork among HEI personnel, consider co-location of key offices
• Co-ordinate the efforts of various offices to support researchers
• A matrix-type structure tends to be most appropriate for TTOs because of its overall capacity to co-ordinate and incentivise the HEI wide interface with business
• The TTO must have a prominent position in the HEI structure and among top management
• Executive Committee to oversee commercialisation activities
• Development of specialists (by TTO activity or by research area)
• Although the functions are of research and technology transfer are related, the two offices have very different goals and should be administratively independent

Source: Bercovitz et al. (2001)

collaboration, specialist offices have developed, which need both co-ordination and leadership at pro-vice-chancellor, vice-rector, or vice president levels. Extensive research by Bercovitz et al. (2001) found that the structure and form taken by the TTO can impact the quality and number of technology transfer outcomes. The growth and importance in the importance of the activities of TTOs necessitate a prominent position in the institution's structure and among the institution's management team. This confers legitimacy on technology transfer activities and facilitates in them being viewed positively and accepted by academics. While there is no optimal structure that a TTO should take, a number of guidelines taken from best practice can facilitate the decision process (see Table 3.3).

3.4.1 Executive Committees

The TTO should be an inherent part of a broader university structure that supports and fosters technology transfer. Relations with research offices are crucial so that the TTO maintains an awareness of the research capabilities of the university and can anticipate and plan for demands on its services. For larger agreements with particular intellectual property rights or financial significance, evidence suggests that TTOs may be supported in its decisions by an executive committee. This committee may be made up of senior members of staff and the top management of the university. The function of the committee would be to provide direction on policy, provide oversight on TTO activities ensuring a balance of research exploitation and academic freedom, while at the same time buffering the TTO in terms of providing support for key decisions. Indeed in his review of best practice Allan (2001) noted that large mega agreements may require regular meetings among senior management of the company and university, as well as a regular exchange at the TTO level.

3.4.2 Co-location of Offices

In further efforts to support the TTO, some universities have encouraged teamwork by co-locating related HEI offices. For example, in the mid-1990 Penn State University decided to cluster the administrative activities that engaged industry into one facility. This has fostered cooperation, rather than competition, for establishing relationships with companies and sharing information. Some universities when further and incorporated their Office of Industry Research Relations and Office of Technology Transfer into a combined Office of Technology Transfer and Industry Research. This is becoming more commonplace given shared service models and co-location approaches being adopted by universities.

3.4.3 Degree of Autonomy

The degree of centralisation of the TTO will generally reflect the culture prevailing at the institution and the size of the institution (Martin 2000). A decentralised approach is natural in the case of a multi-campus university distributed over a number of locations.

3.4.4 Structure by Specialism

The ad hoc development of many TTOs has resulted in them handling a wide range of activities with limited resources. Offices generally cater to a number of different functions such as handling invention disclosures, technology transfer awareness, and networking, supporting business incubator firms and detection of academic entrepreneurship on campus. TTOs have expanded from one-person operations to offices with several specialists working under a director. Each of these specialists tends to focus on one single aspect of technology transfer (see Fig. 3.1). This trend reflects how the role of the TTO has changed significantly over time.

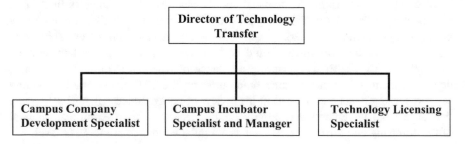

Fig. 3.1 TTO Structure—by specialist activities

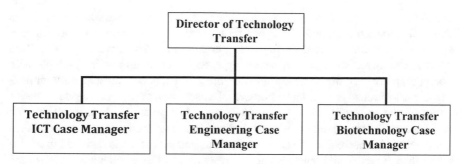

Fig. 3.2   ILO structure—case manager approach

Another model evident from best practice institutions is that TTOs structure their activities in relation to the identified strategic research priorities or technology platforms of the institution (see Key Issue 2). This structure involves specialists in various research areas conducting the full range of technology transfer activities. This 'cradle to grave' approach has been termed 'case management'. The case-management style is commonly practised in many of the TTOs studied by Allan (2001) in his review of best practice (see Fig. 3.2). In this system, one person is responsible for the actions required for a particular case, from disclosure through patenting, and sometimes beyond. This approach offers the advantage of centralising awareness and coordination with all major aspects related to a particular intellectual property. However, this management style requires the talents of very skilled technology transfer professionals with experience in technology transfer processes ranging from invention disclosure through to commercialisation. The breadth and depth of these processes usually leave little time for proactively promoting the opportunity, but this should not be neglected in this type of approach.

In general, both structures suggest a move from multiple role 'one stop entities' towards a more specialised focus reflecting the technology transfer objectives of the university. These structures engender specialists and dedicated attention to specific aspects of the commercialisation process or research areas so that hubs of expertise can be built up. It is more common that you have TTO professionals aligned to university research institutes or centres or particular domain areas.

In the context of such changes, the title of the TTO must reflect the role of the office and its subsequent positioning in the organisational structure. Many offices will have to adapt their structure to fulfil the prerequisite for efficient and successful collaborations with industry.

3.4.5   Structural Options to Overcome Resource Constraints

Best practice in the area of technology transfer usually involves prescriptions, which automatically assume staffing and resources as given. The reality for many smaller TTOs is that such constraints will influence the structure adopted. Due to resource constraints, the role of many TTOs has been generally reactive when it comes to research commercialisation. Evidence from best practice, however, shows that successful technology transfer function stems from a proactive approach. This approach involves auditing and reviewing research programmes for projects with potential, protecting the IP generated from these projects and subsequently marketing them to industry. In a large number of cases, however, resource deficiencies constrain the ability of TTOs to operate to their full potential. As a result of such deficiencies TTO involvement is often limited to serving as an information point on a wide range of areas relating to external relations and assistance is curtailed to advice rather than expertise and only pursing a narrow range of research commercialisation opportunities. Most TTOs are heavily constrained by a limited budget and resource allocation and therefore should search for innovative mechanisms and creative solutions to overcome such deficiencies. Developing a meso-level structure may facilitate co-operation among TTOs thereby reducing resource and expert deficiencies and leveraging economies of scale.

> **Key Success Factor**
> Meso-level structures (e.g. Regional or National) may facilitate co-operation among TTOs thereby reducing resource and expert deficiencies and leveraging economies of scale.

3.4.6 Regional Co-operation: Leveraging Resources and Expertise

In the context of positions of deficiency in expertise, particularly in aspects of commercialisation and knowledge of the patent process, it may serve beneficial to develop more co-operative regional or national mechanisms between TTOs. Such co-operative models maybe a natural progression given that research has suggested that there appears to be a minimum efficient size for running commercialisation activities within HEIs. For example, a 1998 NHS report, for example estimates that average revenues from technology transfer at leading US and UK HEIs are 2.5% of their research income. MIT's revenues are still only 3% of its research income. The same report then estimates that R&D expenditure of £20 million per year is necessary for critical mass in technology transfer, that is to cover the costs of a professional office. Lambert (2003) applied this data to the UK HEI sector, and noted less than 25% of universities would meet this threshold, even though 80% reported trying to run their own operations. This may aid in explaining the limitation diffusion of best practice from TTOs in the United States in the 1990s and 2000s to other countries. Some have argued that partnership between TTOs may provide the essential means for undertaking work that is becoming ever more complex and interdisciplinary (see Brint 2005; Jankowski 1999).

One of the main barriers to commercialisation of university intellectual property discussed in Chap. 2 was the variable quality of TTOs. Clearly for most TTOs, it is not feasible to staff their technology transfer functions with specialists that can work across a number of domain areas. While most HEIs run their own technology transfer operations, the reality is that it is only an elite few that have sufficient levels of research activity to justify the required investments to build high-quality offices with substantial expertise. However, it may be feasible to have specialists in particular areas if there was a system in place that allowed the technology transfer specialists to work across different universities. Variants of such co-operative approaches are not uncommon. For example in Denmark, Larson (2000) noted that in the Copenhagen area four universities had formed an amalgamation of biotech competencies in a type of joint property, also open to participation from private firms. On the same site, there was a construction of a research park as an integrated part of the merger. More current examples come through national research programmes that require universities to form research consortia to pursue research in domains of national economic and industrial importance. More contemporary examples of this approach include the Industrial Strategy in the UK and Programme for Research in the Third Level Institutions (PRTLI Cycles 1–5) in Ireland.

Graff et al. (2002: 28) note 'it is reasonable to assume that the marketing of HEI research would benefit from economies of scale, meaning that, at least to an extent, the larger the represented research base, and thus the larger the portfolio of new technologies made available to license, the lower the marketing costs and the higher the probability of successfully selling licenses'. These conditions could be achieved if several universities used a single TTO to market their innovations, a move that would increase the variety and depth of technologies available and thus increase the attractiveness of the combined research base to potential business customers. The establishment of a middleman organisation can, therefore, be a 'viable alternative'. Meyer (2003) notes that: 'for campuses with a technology transfer office but with insufficient staff it is possible to outsource some of the activities-licensing professionals assisting the HEIs with invention evaluation for IP and commercial potential as well as marketing of inventions and license negotiation'. In more recent times that has been a growth of virtual TTOs who in essence work with established TTOs to realise research commercialisation through their channels and networks, and in doing so share in the rewards.

3.4.7 A Co-operative Model for TTOs: Shared Services

The potential benefit of such a system of co-operation is twofold. Firstly, it provides an on-the-ground presence in each university for technology transfer issues. Stimulants for technology transfer identified in Chap. 2 included good relationships with researchers and academics coupled with knowledge of the research activities of an institution. Such knowledge is crucial in developing trust with academics and for promoting the merits of the commercialisation process. Secondly, particularly

Table 3.4 A co-operative model of TTOs

Shared services could include	HEI-based services
Licencing negotiation	Raising awareness of IP issues
Market research for new technologies	Negotiating collaborative research contracts
IP marketing	Reach-out to business
IP management	Consultancy contracts
Spin-out creation	Local contact
Legal expertise	

Source: Adopted from Lambert (2003: 56)

important in the context of the aforementioned expertise deficiencies, the model provides for specialist expertise in particular fields and industries to be shared. Such an approach is especially beneficial for less research-intensive campus without a technology transfer office on the campus. A system-level person could be designated to serve several campuses. The model would also enable universities to share costs in terms of the legislative details of the patenting process. Indeed, should universities co-operate they would have greater bargaining power for negotiating deals with the larger law firms. As well as providing economies of scale co-operative mechanisms would provide scope for mutual learning and diffusion of best practice.

A shared service model does not involve the closing down of technology transfer offices in any university (technology transfer staff need close links with researchers to gain their support for commercialisation, to find out about new technologies in the research pipeline, and to locate potential industry customers in their specialist field) but rather it highlights the benefits of a *complementary* structure at a more macro level drawing on the joint resources of a number of institutions. The Lambert Report (2003) provides some useful insights into the potential mechanisms and structure of such a co-operative body (Table 3.4).

A variation of this model has been adopted in an Irish context with the support of the Irish government through two cycles of strengthening initiatives that support building TTO capacity through the third-level sector as well as creating Knowledge Transfer Ireland.[4] In comparison study of university research commercialisation in the United States, Ireland, and New Zealand Geoghegan et al. (2015) found TTO scale and university leadership were critical determinants to success.

The broadening of the range of issues confronted by TTOs clearly creates a challenge. As demand for service grows, the resources and capacity of TTOs can be stretched. These pressures coupled with the difficulty attracting and compensating technology transfer professionals suggest that due consideration should be given to collaborative structures or even reducing the focus and services offered to university communities.

---

[4]See https://www.knowledgetransferireland.com

3.5 Expertise and Resources

> **Key Success Factor**
> Strong leadership, investment in the skills, and capabilities of technology transfer professionals and configuring a sustainable resource base to support services are critical factors for TTO effectiveness.

A major challenge for TTOs is sourcing skilled professionals. Deals have become more complex and TTO staff must master many competencies. Apart from solid knowledge in the specific scientific fields, TTO staff must have legal and economic competence to judge whether inventions are patentable or not, marketing, innovation, and entrepreneurial and business acumen skills in order to find commercial partners and finally, negotiation and social skills to be able to finalise a good agreement (Karlsson 2004). TTO activity, therefore, requires both technical and boundary-spanning skills. Overall technology transfer staff need close links with researchers to gain their support for commercialisation, to find out about new technologies in the research pipeline, and to locate potential industry customers in their specialist field. Increased emphasis has been placed on marketing and negotiation skills as TTO staff can be successful in eliciting an invention disclosure, patenting it but then suffer difficulties in licencing it to business. Barriers to effective technology transfer are often a function of the deficiency in skills and abilities of the TTO for coordinating activities and commercialising research (Graff et al. 2002; Sanchez and Tejedor 1995; Scott et al. 2001). The varying demands placed on TTO staff highlight the requirement for strategic staffing based on expertise and the need for continuous training and development (Lambert 2003).

3.5.1 Strong Leadership

Clearly having appropriately trained and skilled staff is an essential prerequisite to for an effective TTO. Best practice institutions go to great lengths and offer competitive remuneration packages to secure the best technology transfer talent. TTOs, therefore, require sufficient budget allocation to support these initiatives and to invest in commercialisation efforts. The recruitment of appropriate and capable leadership is of crucial importance. Some leadership criteria drawn from best practice include:

- A strong record of leadership, management, and administration of a sizable unit, with demonstrated abilities to develop and articulate a vision for an organisation and to plan strategically.
- An in-depth knowledge of universities, the academic, investor, industry communities, and of their corresponding cultures.

- In-depth knowledge and experience of intellectual property issues, the principles of licencing and spin-off company creation, as a means to carry out technology transfer objectives.
- Outstanding interpersonal skills and a proven collaborative management approach with the ability to consult widely, internally and externally, and to inspire confidence.
- A commitment to the broader goals of the university, and to building bridges with units of the university and between the university, industry, and government.
- A strong record of mobilising resources and securing funds from a range of sources.

Research has shown that staffing within TTOs can facilitate in explaining why some HEIs are more proficient than others in managing intellectual property. According to Parker and Zilberman (1993), TTOs usually hire either a mix of scientists and lawyers or a mix of scientists and entrepreneurs/business people. In the former case, legal functions, such as the adjudication of disputes involving intellectual property rights and the negotiation of licencing agreements, are performed in-house. In the latter case, such functions are usually outsourced. Parker and Zilberman (1993) hypothesise that the entrepreneur/business model for TTOs may be more conducive to helping scientists form their own start-ups. It also seems reasonable to assume that TTOs staffed in this manner would be more effective in the marketing phase of university–industry technology transfer. Siegel et al. (2003a, b) found that a substantial percentage of managers suggested that universities hire more licencing professionals with stronger marketing and business skills. Furthermore in the entrepreneur/business model legal expertise may be shared among a number of TTOs. Siegel et al. (2003a, b) research is again instructive as their findings imply that spending more on external lawyers reduces the number of licencing agreements, but increases licencing revenue.

3.5.2 Skilled and Flexible Personnel

In order to keep TTO staff up to date with university-based technology transfer, it is necessary to provide continuous training and professional development. Overall technology transfer requires well-skilled and flexible personnel. Visible commitment from top management is essential in attracting the best talent to work at a university TTO (Den Hertog et al. 2003: 99). As technologies become more complex and industry becomes more knowledge-driven technology transfer specialists will become more important. In addition, to qualifications in specific technological fields, greater weight has been given to the skills necessary to carry out the job. In the case of a technology transfer specialist, marketing, communications, and negotiation skills are likely to be just as important as having an in-depth understanding of the science behind the innovation. The continuing involvement of the researcher may provide this scientific understanding. Most TTOs surveyed by the OECD stressed the importance of the technology transfer officers' informal relations. This may suggest why in many cases it is recommended that staff hired for TTOs have industry experience and hence an understanding of the industry but equally important a network of contacts and linkages (OECD 2003). Other mechanisms of staff development include periodic staff meetings and formal and informal discussions of cases under management by TTO staff. Sharing of this collective experience provides an array of how-to advice and extremely valuable insight, especially for junior members of TTO teams.

3.5.3 Integration of the Strategy of the TTO and Its HR Strategy

The human resource strategy of TTOs and competencies they emphasis and nurture must be integrated and supportive of the objectives as laid out in the strategic plan. Consideration of, and investment in, the HR strategies of the TTO will serve as a further indicator of the commitment by the university and the TTO itself to technology transfer and commercialisation activities. Induction programmes in TTO might involve placements in other TTOs as well as shadowing colleagues undertaking other duties, so as to get a deeper understanding of their role and how it integrates with others.

3.5.4 Resources

Resource endowments, particularly financial, to facilitate the process of technology transfer are essential for a more proactive approach to technology transfer activities. In some contexts, TTO involvement is often limited to serving as an information point on a wide range of areas relating to external relations and assistance is curtailed to advice and facilitation. Clear consideration of the skills required and matching of HR strategies with TTO objectives should aid in reducing such deficiencies.

Resource Sharing
In cases where such resource investments are not possible in-house, i.e. are constrained by the size or research base of the university, consideration should be given to structures that facilitate the sharing of expertise, especially expensive legislative know-how. Marketing and business skills could also be obtained from interaction and involvement with business faculty members. Indeed one of the most underutilised resources, particularly in working with potential campus companies, is the facilities and expertise of the business faculty. Academics in business faculties have a potentially valuable contribution to make to spin-off company projects as part of the project work for suitable courses, such as MBA or MSc programmes.

3.6   Conclusions

A key takeaway from this chapter is that TTOs need to take a strategic perspective that is aligned with its current context and in doing so they need to build an organisational structure and a resource base that allows for future scaling of their activities. Recruiting and investing experienced professional staff within TTOs is critical. In summary, the key success factors for the strategic aspects of our business model framework are:

- Development of a strategic plan, mission statement, and strategic priorities.
- Focused more on action as opposed to facilitation.
- There is no one best way to structure TTO activities.
- Structure will be heavily shaped by institutional factors and strategic focus.
- Meso-level structures (e.g. Regional or National) may facilitate co-operation among TTOs thereby reducing resource and expert deficiencies and leveraging economies of scale.
- Strong leadership, investment in the skills and capabilities of technology transfer professionals and configuring a sustainable resource base to support services are critical factors for TTO effectiveness.

References

Allan MF (2001) A review of best practices in university technology licensing offices. J Assoc Univ Technol Manag 13(1):57–69

Aragonés-Beltrán P, Poveda-Bautista R, Jiménez-Sáez F (2017) An in-depth analysis of a TTO's objectives alignment within the university strategy: an ANP-based approach. J Eng Technol Manag 44:19–43

Beesley LG (2003) Science policy in changing times: are governments poised to take full advantage of an institution in transition? Res Policy 32(8):1519–1531

Bercovitz J, Feldman M, Feller I, Burton R (2001) Organizational structure as a determinant of academic patent and licensing behavior: an exploratory study of Duke, Johns Hopkins, and Pennsylvania State Universities. J Technol Transf 26(1–2):21–35

Brint S (2005) Creating the future: 'New directions' in American research universities. Minerva 43 (1):23–50

Chang YC, Yang PY, Chen MH (2009) The determinants of academic research commercial performance: towards an organizational ambidexterity perspective. Res Policy 38(6):936–946

Cunningham JA, Golden W (2015) National innovation system of Ireland. Wiley Encycl Manag 13:1–14

Cunningham J, Dolan B, Kelly D, Young C (2015) Medical device sectoral overview. In: Galway city and county economic and industrial baseline study. The Whitaker Institute, Galway

Den Hertog P, Gering T, Cervantes M (2003) Introduction and overview, chap. 4. In: OECD (ed) Turning business into science: patenting and licensing at public research organisations'. OECD, Paris

Fassin Y (2000) The strategic role of university-industry liaison offices. J Res Adm 1(2):31–42

Geoghegan W (2017) University/industry collaboration and the development of firm-specific innovative capabilities. In: The world scientific reference on entrepreneurship: vol. 1: Entrepreneurial universities technology and knowledge transfer. World Scientific, New Jersey, pp 185–222

Geoghegan W, Pontikakis D (2008) From ivory tower to factory floor? How universities are changing to meet the needs of industry. Sci Public Policy 35(7):462–474

Geoghegan W, O'Kane C, Fitzgerald C (2015) Technology transfer offices as a nexus within the triple helix: the progression of the university's role. Int J Technol Manag 68(3–4):255–277

Giblin M (2011) Managing the global–local dimensions of clusters and the role of "lead" organizations: the contrasting cases of the software and medical technology clusters in the West of Ireland. Eur Plan Stud 19(1):23–42

Giblin M, Ryan P (2015) Anchor, incumbent and late entry MNEs as propellents of technology cluster evolution. Ind Innov 22(7):553–574

Goel RK, Göktepe-Hultén D (2018) What drives academic patentees to bypass TTOs? Evidence from a large public research organisation. J Technol Transf 43(1):240–258

Göktepe-Hultén D (2010) University-industry technology transfer: who needs TTOs? Int J Technol Transf Commer 9(1):40–52

González-Pernía JL, Kuechle G, Peña-Legazkue I (2013) An assessment of the determinants of university technology transfer. Econ Dev Q 27(1):6–17

Graff G, Heiman A, Zilberman D, Castillo F, Parker D (2002) Universities, technology transfer and industrial R&D. Econ Soc Issues Agric Biotechnol:93–117

Green R, Cunningham J, Duggan I, Giblin M, Moroney M, Smyth L (2001) The boundaryless cluster: information and communications technology in Ireland. Innov Clust:47–64

Hernes G, Martin M (2001) Management of university-industry linkages. Policy forum no. 11. Proceedings from the policy forum held at the IIEP (Paris, June 1–2, 2000). International Institute for Educational Planning, Paris

Jankowski JE (1999) Trends in academic research spending, alliances, and commercialization. J Technol Transf 24(1):55–68

Jones-Evans D, Klofsten M, Andersson E, Pandya D (1999b) Creating a bridge between university and industry in small European countries: the role of the industrial Liaison Office. R&D Manag 29(1):47–56

Karlsson M (2004) Commercialization of research results in the united states an overview of federal and academic technology transfer. ITPS, Swedish Institute for Growth Policy Studies

Lambert R (2003) Lambert review of business-university collaboration. Final report, December

Larson, K (2000) In response to Moshe vigdor paper. In: Hermes G, Martin M (eds) Management of university-industry linkages. Policy forum, no. 11 UNESCO/International Institute for Educational Planning, Paris

Leischnig A, Geigenmüller A (2020) Examining alliance management capabilities in university-industry collaboration. J Technol Transf 45(1):9–30

Lundvall BA (2002) The university in the learning economy. DRUID working paper (02-06). Aalborg University

Martin M (2000) Management of university-industry linkages. Policy forum no.11 UNESCO/International Institute for Educational Planning, Paris

Meyer JF (2003) Strengthening the technology capabilities of Louisiana universities, Pappas & Associates report, Prepared for the Louisiana Department of Economic Development, June 30

Mosey S, Lockett A, Westhead P (2006) Creating network bridges for university technology transfer: the Medici fellowship programme. Tech Anal Strat Manag 18(1):71–91

Muscio A (2010) What drives the university use of technology transfer offices? Evidence from Italy. J Technol Transf 35(2):181–202

OECD (2002) Interim results of the TIP project on the strategic use of IPRs at PROs. Working document, June

OECD (2003) Turning business into science: patenting and licensing at public research organisations. OECD, Paris

Parker DD, Zilberman D (1993) University technology transfers: impacts on local and US economies. Contemp Econ Policy 11(2):87–99

Sanchez AM, Tejedor ACP (1995) University-industry relationships in peripheral regions: the case of Aragon in Spain. Technovation 15(10):613–625

Scott A, Steyn G, Geuna A, Brusoni S, Steinmueller E (2001) The economic returns to basic research and the benefits of university-industry relationships a literature review and update of findings. Report for the office of Science and Technology by SPRU

Shattock, M. (2001). In what way do changing university–industry relations affect the management of higher education institutions. G. Hernes and M. Martin, Management of University-industry Linkages, International Institute of Educational Planning, UNESCO, Paris

Siegel DS, Waldman DA, Atwater LE, Link AN (2003a) Commercial knowledge transfers from universities to firms: improving the effectiveness of university–industry collaboration. J High Technol Managem Res 14(1):111–133

Siegel DS, Waldman D, Link A (2003b) Assessing the impact of organizational practices on the relative productivity of university technology transfer offices: an exploratory study. Res Policy 32(1):27–48

Yuan C, Li Y, Vlas CO, Peng MW (2018) Dynamic capabilities, subnational environment, and university technology transfer. Strateg Organ 16(1):35–60

Chapter 4
Business Model Framework: Operational Considerations

Abstract There are many operational considerations in establishing and running an effective technology transfer office. For business model framework operational considerations, we focus on policies and procedures, technology transfer mechanisms, evaluation, and outcomes. The culture and ethos for commercialisation and researcher motivation is also explored and how this can shape technology transfer activities. We conclude with some lessons that focus on taking a strategic approach using key issues, factors for success, and facilitating factors.

Keywords Technology transfer · TTOs · Researcher motivation · Culture · Policies · Mechanisms

4.1 Introduction

In this chapter, we continue our focus on the business model framework by focusing on operational considerations. In particular, we focus on the different aspects of policies and procedures, technology transfer mechanisms, evaluation and outcomes before finally concluding by examining culture and commercialization and researcher motivation.

4.2 Policies and Procedures

The policies and procedures TTOs use impacts on their operational effectiveness and faculty perceptions. These are necessary to support effective intellectual property protection. However, if the compliance with the policies and procedure adds significantly to the faculty workload or is perceived as too cumbersome this result in the TTO being bypassed or the researchers not fully engaging with the TTO. In essence this creates barriers between the TTO and researchers (see Cunningham et al. 2014; O'Reilly and Cunningham 2017; O'Kane et al. 2017). Therefore, TTOs need to carefully plan their policies and procedures from an operational perspective

© The Author(s), under exclusive licence to Springer Nature Switzerland AG 2020

J. A. Cunningham et al., *Effective Technology Transfer Offices*, SpringerBriefs in Business, https://doi.org/10.1007/978-3-030-41946-2_4

from processing invention disclosures to undertaking the necessary activities to support different technology transfer mechanisms.

4.2.1 User-Friendly Policy and Procedures

In order to encourage commercialisation activities, the TTO needs to develop concise, easy to understand transparent policies to ensure that procedures are maintained and followed. With respect to policies and practices, user-friendly means that they meet the norm of peer institutions and industry finds them attractive and wants to do business with the university. A key issue for many researchers, particularly young researchers, is a lack of understanding of intellectual property issues and the array of technology transfer mechanisms. There is a need to develop user-friendly information packs to all researchers to help build their awareness. These could include infographics, short animated video clips of commercialisation process and activities of the TTO. A user-friendly approach could also involve:

- Providing adequate outreach/educational presentations and web materials
- Demonstrating the skill, ability, and willingness to negotiate with all bodies
- Being professional and factual in explaining why an invention will not be patented and/or marketed. At least 70% of the inventions received by the technology transfer office will not be licenced. It is important to take the time to explain why the marketplace will not or has not decided to licence an invention (Meyer 2003: 11).

Other user-friendly approaches utilised in other instructions include the development of researcher kits for each type of agreement. These researcher kits include an explanation of when to use the agreement, a lay-language outline of the agreement terms; a research questionnaire to identify the relevant issues; and a step-by-step outline of the process. In order to ensure that policies are user friendly, the development process could involve a review or trail run by a focus group of faculty researchers. Policies should also be systematically reviewed to ensure that they are developed and/or updated at both the system and the campus level, with each addition subsequently examined for user-friendly wording. This should involve seeking inputs from faculty, industry, and other relevant stakeholders to ensure that the policies and procedures are fit for purpose and are user friendly.

4.2.2 Disclosure Process

To be effective, TTOs need to have an efficient inventions disclosure process. Resources should be allocated so that appraisals are conducted quickly and there is a separation of 'highly promising' from 'promising' submissions. This helps establish legitimacy and the political value of the office. Knowledge of an efficient and fair system encourages researchers to disclose their inventions. The disclosure process in TTOs should, therefore, ensure a specified, rapid turnaround time for disclosure assessments, based on a clear understanding of the proposed technologies. Essentially evidence from best practice indicates the importance of timely responses to invention disclosures, communicating clearly with researchers and leveraging expertise to ensure the disclosures chosen are the ones with the most potential opportunity are critical features of an effective disclosure process. Some issues to consider:

Pre-disclosure process—Meeting between researcher and TTO to discuss and scope potential.

Invention disclosure—Consider template design and information requirements so that a thorough consideration can be undertaken and it can be shared with relevant expertise inside and outside the TTO to reach a final determination.

Feedback—Define a service-level agreement timeframe for feedback—20 or 30 working days and also how feedback is provided to a researcher including next steps.

4.2.3   Negotiating Agreements

In addition to HEI licencing policies, premature definition and valuation of intellectual property can become an obstacle at the initiation stage of a collaborative project. Granting the industry partner the right of first refusal to negotiate an exclusive licence is one commonly used practice to delay concrete negotiations until the commercial value of an invention is easier to assess. When the extent of commercialisation activities is of sufficient magnitude or in the case of larger university–industry research relationships, the use of master contracts or templates should be considered. These master agreements have grown and such templates do provide useful parameters for negotiation or for how to proceed with exploiting new knowledge. Furthermore, standardisation reduces the risk of error or litigation as a result of certain clauses. Such contracts should be continuously reviewed to ensure that they keep up with the most recent developments. Universities also should consider developing model agreements and ensure that the terms do not unduly disadvantage small- and medium-sized companies.

Confidentiality agreements, when necessary should be signed by the industry partner, the university, and the researchers involved. The industry partner and the university must take responsibility for safeguarding confidential information. Publication delays to protect intellectual property rights should generally be no longer than 60–90 days. Any publication delays should be carefully monitored both to preserve academic freedom and to protect against any early disclosure that might invalidate patent claims. Clear procedures with regard to this will facilitate in easing discomforts that some academics will naturally have in terms of the commercialisation of their research results.

4.2.4 Conflicts of Interest

Conflict of interests as defined by the OECD (2002) refers to a situation where: 'an official's private interests—not necessarily limited to his/her financial interests—and the official duties in his/her public function are in conflict (actual conflict of interest) could come into conflict or could reasonably appear to be in conflict (potential conflict of interest)'. Conflicts of interest can exist among faculty, between the TTO and researchers and also among non-faculty such as graduate students.

Even when expectations are well managed some forms of conflict can become apparent. Even the perception of possible conflicts of interest could prove to be extremely damaging to the reputation of the university and the industry partner concerned. Perceptions of a conflict of interest can weaken public trust—a particular concern for university research, which is heavily depend on government research funding. Conflicts of commitment are generally defined as anything that might interfere with a faculty member's full-time duties. Many universities have formal policies limiting the amount of time that a faculty member can spend engaged in outside activities.

A natural outgrowth of an increased focus on business–university interactions is an increased likelihood of conflicts of interests arising. The real issue then becomes what methods should be put in place to resolve conflicts once they come about since their emergence is a predictable part of the process of change (see Mathieu 2012). TTOs must adopt clear policies to avoid potential conflicts of interest and have developed clear guidelines to deal with conflicts of interests as they arise. Guidelines provide templates on how to avoid or manage potential conflict of interests between the researcher's obligation and the more entrepreneurial activities such as patenting inventions, carrying out contract research, or working in a start-up company or spin-off. Guidelines for dealing with individual conflicts of interest vary but there are common elements such as disclosure of all relevant financial interests and activities outside the framework of employment or disclosure of financial interests to non-government sponsored research. As university officials, researchers, and the companies with which they collaborate study these conflict-of-interest issues, there are a number of basic principles that should be remembered and which serve as parameters to the creation of guidelines. These include:

- The core values of academic freedom must be maintained.
- Industry funding cannot, and should not, be viewed as a substitute for adequate, long-term public financing of basic scientific research.
- University and companies should seek transparency, clarity, and consistency.

The precise mechanisms for managing a conflict usually depend on the details of each case. Options can include divesting troublesome assets, ending consulting arrangements, withdrawing the researcher from the project, independent review, and disclosing significant financial assets in any published report on the research. A useful strategy for preventing potential conflicts involves ongoing education, aimed both at faculty members and at graduate students who hope to become practicing scientists (RCI 2003).

4.2.5   Sponsored Research

Often conflicts of interests arise as companies request background rights to sponsored research projects. Companies have legitimate reasons for requesting background rights to sponsored projects and, as part of their due diligence, should assist universities in locating potential conflicts. HEIs have legitimate reasons for not providing background rights, but they should make a strong effort to do so when appropriate and feasible. HEIs should closely consult with faculty and confirm that all contractual obligations can be met before signing binding agreements (RCI 2003). Detailed consideration and consultation can reduce the likelihood of conflict and facilitate in resolving it should it arise.

Other conflicts of interest may arise with regard to the misuse of student time. For example, when a faculty member holds an equity stake in a company that sponsors university research and has graduate students working on that research, potential conflicts can arise. In order to address such issues, universities have developed policies to deal realistically with these issues. Although some faculty members may wish to minimise involvement in collaborations by the universities' research administration and departments, others assert that open communication with these bodies can prevent instances, as well as suspicion and misunderstanding about such arrangements. Some universities and departments simply do not allow students to become regular or part-time employees of research sponsors.

4.2.6   Intellectual Property Rights

The activities of TTOs are greatly facilitated by active protection and management of intellectual property. Due consideration to intellectual property-related issues can simplify negotiations. Policies that help with this are outlined in Table 4.1.

4.2.7   Clear Codified Intellectual Property Rights

One of the barriers to commercialisation cited frequently is a lack of clarification over intellectual property rights. Such confusion can result from a lack of transparency over procedures and regulations. Clearly, when establishing collaborative research partnerships, it is important to determine at the outset the ownership and exploitation rights for any intellectual property that may be generated. Business and university both report that negotiations on the terms and conditions of IP ownership

Table 4.1 Intellectual property protocol main features

Intellectual Property protocol main features could include:
– The common starting point for negotiations on research collaboration terms should be that universities own any resulting intellectual property with industry free to negotiate licence terms to exploit it
• But if the industry makes a significant contribution it could own the intellectual property
• Whoever owns the intellectual property, the following conditions need to be met:
1. The university is not restricted in its future research capability.
2. All applications of the IP are developed by the company in a timely manner.
3. The substantive results of the research are published within an agreed period.
• On all other terms, the protocol should recommend flexibility where possible to help ensure that the deal is completed.
• The Funding Councils and Research Councils should require universities to apply the protocol in research collaborations involving funding from any of the Councils.

and exploitation can be extremely lengthy and costly. Research continuously refers to the importance of clear codified guidelines, particularly in the areas of ownership of intellectual property, equity shares, and distribution of royalties.

4.2.8 Harmonisation of Intellectual Property Rights

TTOs should be involved in efforts to encourage further harmonisation—or at least compatibility—of national rules regarding intellectual property rights, which may also, in turn, facilitate collaborative research by reducing transaction costs. In many countries, government agencies have been heavily involved in promoting such harmonisation and diffusing knowledge and expertise in respect of intellectual property rights in the form of raising awareness events dealing with intellectual property along with the benefits of university–industry collaboration at academic and industry events (see Cunningham et al. 2020). Also, this has led to many countries harmonising process and contract templates to reduce the burdens for industry and universities. Knowledge Transfer Ireland is an example how this can be achieved that benefits all stakeholders and provides the clarity and consistency that is needed.

4.2.9 Distribution of Intellectual Property Benefits

Guidelines should clearly outline the reward system of shared benefits. Ideally, rewards should be weighted positively towards the inventor to act as an incentive for research commercialisation. This is particularly the case for universities trying to develop and attract attention and interest towards commercialisation activities. Clearly, it is critical to have distribution policies in place prior to the commencing of the commercialisation process and before the research concept becomes valuable, at which stage, reaching an agreement can become a very complicated and legal business. The most successful policies are presented in a user-friendly format and described at orientation programmes. In addition, intellectual property rights policies should be changed to reflect the fact that innovations are often developed as a result of a team effort and should be shared between different members of the team.

4.2.10 Perceived Publishing Constraints

Companies should secure rights to the intellectual property they want to commercialise, but it is also important that any deal on intellectual property should not unreasonably constrain the university from publishing the results in a timely fashion, from doing further research in the same area, or from developing other applications of the same intellectual property in different fields of use. It follows from these points that there should be as much flexibility as possible in the distribution of intellectual property rights between universities and industry.

4.2.11 Flexibility in Terms of Various Intellectual Property Policies

It has been noted that the maximum creative use of IP allows the full economic potential of research collaboration to be unlocked (Sanchez and Tejedor 1995). If business negotiates full ownership of the intellectual property with strong restrictions on university use, this may reduce the total economic impact of the intellectual property in the future. Therefore, it is important that there is flexibility in the distribution of IP rights between universities and industry. Increased flexibility in the use of Intellectual Property prevents from being locked up in a way that limits its exploitation across as wide a range of areas as possible.

IP Protocol
Research on best practice suggests the best way to meet these objectives is to introduce an intellectual property protocol (see Allan 2001; Lambert 2003). This would provide simple ground rules for negotiations and encourage the flexible use of the intellectual property by both universities and business. In most cases, universities make a significant contribution to collaborations, so the default position should be that they own the IP. Companies, however, may own the IP whenever their contribution is significant. Sample or previous contracts could also be used as templates to facilitate in reaching a voluntary agreement between universities and business.

Joint Research Contracts and IP
It is much more difficult to agree on the ownership of IP in research projects that have been funded by both universities and industry. Some business funding for

university research is in this form. IP ownership is often strongly contested in these research collaborations, as the sponsors have different interests in the right to exploit and use the IP. The rationale guiding universities' demand for the IP is that ownership is required to ensure that their future research is not held back. Industry often argues that it needs ownership to protect the investment, which will be required to develop the IP into a commercial product. Although ownership and control of intellectual property resulting from a collaboration must be decided by the collaboration partners, it usually will be appropriate for the university to retain ownership. Both parties should remain flexible during negotiations, and the key measure should be whether the corporate partner has the ability to commercialise the fruits of the research to the benefit of the public. Universities should update their copyright policies to allow industry sponsors to be granted licencing terms on a basis similar to that provided with patents. Occasionally the resolution mechanism is that intellectual property ownership should be proportionate: the party that makes the biggest contribution (intellectual as well as financial) should have first rights on the IP ownership.

Universities taking a restrictive approach to licencing and placing too high a value on their intellectual property contributions can hamper some IP negotiations. Also, university boards of trustees may see technology transfer activities more as a revenue source than as a component of the HEI's public responsibility to assist in commercialising research results (see Siegel et al. 2003; Scott et al. 2001). This attitude can raise barriers to negotiations that actually reduce revenue in the long term. Given that only a small percentage of university-generated inventions produce significant revenue, some participants likened the strong emphasis on protecting proprietary rights of some universities to 'buying lottery tickets'. Universities would, therefore, benefit from being less aggressive in exercising their intellectual property rights (Siegel et al. 2003). The role of intellectual property in the innovation process varies by field. In some cases, universities do not seek patents on their inventions unless an industry licencee has been identified. Allan (2001) notes that this approach is much more likely to facilitate commercialisation rather than a blanket policy of not patenting inventions outside the life sciences, which is evident at some universities.

4.3   Technology Transfer Mechanisms

From Chap. 3 and our discussions on strategic plans, we noted that TTOs should pursue a focus and planned approach to commercialisation and technology transfer. TTOs need to be clear what is their scope of activities and as a result what services that are going to provide to the university community they serve and to industry. Consideration should be given to leverage research and resources from the previously identified technology platforms. Research in this area stresses that TTOs and their activities are distinct products of their associated institutions' history and development path. Each TTO must, therefore, choose a mechanism for technology transfer that suits their specific needs and circumstances.

> **Key Issue**
> There is no one best mechanism for technology transfer each TTO must choose mechanisms for technology transfer specific to their individual needs and circumstances.

4.3.1   Spin-Offs and Start-Ups

There is burgeoning research on university spin-off (see Benneworth and Charles 2005; Miranda et al. 2018; Mustar et al. 2008; Link and Scott 2017; Rasmussen et al. 2011). Spin-offs and start-ups are seen to be a direct manifestation of technology transfer and therefore prove popular technology transfer mechanisms for institutions initiating or developing mechanisms for technology transfer. Elements of good practice for developing spin-offs include:

- Provide clear and transparent guidelines to all university staff wishing to set up a spin-off company, so that the prospective entrepreneurs have a solid framework before they make their judgement about whether to proceed with setting up the company and how this should be done.
- Proactively provide information, contacts, and support on how to establish a spin-off firm.
- If the university owns the IP which protects the innovation or technology, which forms the basis for the company, it needs to jointly discuss and decide at the outset with the academic inventors or prospective entrepreneurs whether setting up a spin-off company is the best option.
- Prospective entrepreneurs should be allowed to remain on a part time research or teaching position with the university if they so desire. There should be a clear demarcation of responsibilities and practices between the two jobs.
- Provision of advice to prospective academic entrepreneurs on what is required to effectively operate and run the company and what additional manpower expertise may be needed to be recruited to fill the gaps.
- Some support to help business models and plans to define clear commercial goals for the new company as well as addressing market research and sales strategies.

It is important to have appropriate support mechanisms in terms of allowing academic staff leave to work on such ventures. Starting a new firm does not necessarily imply that they leave their academic position and take up a consulting position or a board position in the new firm.

4.3.2   The Risks of Spin-Outs

The start-up is only one of the exploitation mechanisms at the TTOs disposal for obtaining maximum benefits from its research results. Indeed, there is a strong view from both business and universities that the balance of commercialisation activities has moved too far towards spin-outs, with too little licencing and too many unsustainable spin-outs. The Lambert Report (2003) noted that high spin-out rates come at a cost to licencing. While spin-out activities of TTOs have led to some successful companies being created an ongoing danger is too many spin-outs created can result in low-quality and less commercially viable firms. TTOs need to get this balance right and do what is in their control to reduce any barriers to spin-out activities

In getting this balance TTOs, therefore, need to carefully assess the likelihood of a spin-off being successful and assess the resources and time that can be reasonably invested in encouraging such activities. While an attractive method of technology transfer, it is often the case that resource constraints and a lack of expertise makes these activities unsustainable. Therefore, some TTOs have supported or have been directly involved in setting up incubator or accelerator programmes (Bergek and Norrman 2008; Binsawad et al. 2019; Rothaermel and Thursby 2005; Schwartz and Hornych 2010).

4.3.3   Licencing

Licencing has become one of the most popular avenues for research commercialisation activities and been the focus of much research attention (see Thursby and Thursby 2002, 2007; Kim et al. 2019; Thompson et al. 2018). Licencing is less resource intensive than spin-outs—in terms of both people and funding—and has a higher probability of getting technology to market. It is often the quickest and most successful way of transferring IP to industry, and has the advantage of using existing business expertise rather than building this from scratch. However, TTO should be selective in relation to what intellectual property it will protect through the patenting process that forms the basis of licencing agreements.

In order to successfully develop licencing capabilities, TTOs will need to have greater capacity and resources for managing IP. Critical in this respect is government support and funding to increase the availability of proof-of-concept funding. Proof-of-concept funding is used to establish whether a new technology is commercially viable or not. It is the first stage in transferring IP to the market. Proof-of-concept funding is needed for both licencing and spinning out. TTOs should therefore actively promote the importance of proof-of-concept funding and work with local development agencies and other TTOs to ensure adequate resources are allocated to these activities.

4.3.4 Portfolio Approach: Pursuing Many Deals

One approach utilised by the US institutions has been termed portfolio management, whereby TTOs pursue a number of commercialisation channels to minimise the risk of being exposed by any particular one. Statistics from the Associate of University Technology Managers (2002) established a positive correlation between royalty income and the number of active licences at an academic institution. Accordingly, many institutions adopt a portfolio approach in order to pursue many opportunities simultaneously, aiming for success collectively. The portfolio approach requires that TTOs pursue many agreements for their technologies. Linked with such an approach is the prioritisation of daily activities to make progress on signing deals. Such prioritisation should be made in the context of the TTOs' overall strategic objectives. Given that one in nine deals pays out royalties, the success of the TTO will depend heavily on the number of deals made.

A portfolio approach serves to spread the risk of commercialisation activities. Based on priorities as laid out in the strategic plan and based on each institution's specific contingencies, a number of different channels will be pursued to facilitate in getting the deal done. Such an approach is often more realistic in the case of resource-constrained institutions. While some universities invest in start-up firms or accept equity in lieu of royalties on university-held patents, this raises concerns that they might become beholden to a company in which they have a financial stake. Ultimately, developing multiple funding sources can help protect universities from becoming indebted to any one entity.

4.4 Evaluation and Outcomes

A crucial dimension of developing a more strategic approach to TTOs' activities involve the introduction of an evaluation mechanism to ensure that the office is conducting and prioritising activities as set out in its strategic objectives. While TTO measurement and evaluation have grown in the early stages of office development can be neglected as part of the overall organisation structure and design of an office. Taking stock of organisational practices in university management will be useful in many regards. Given the embryonic nature of the TTOs generally, there is a need to document practices being conducted. University administrators and policymakers often express a strong interest in benchmarking their IP management practices relative to peer institutions. While such benchmarking might be useful, the danger is that such analysis does contribute and not dominate how a TTO evolves within their own contextual environment. Documentation of activities, performance, and success stories should be disseminated to various stakeholders. In a university context, they serve to legitimise and make transparent the role of the TTO as part of the general university structure. Crucially the introduction of indicators cannot be seen to be proposed merely for their role as performance metrics but must be promoted to staff for the way in which the indicators will facilitate university administrators in the support and management of activities. Clear linkages of the objectives to broader university-wide objectives can promote co-operation and a sense of common purpose.

4.4.1   Holistic Approach

There is no one model for a TTO and this is very much conditioned by context. Learning can be facilitated by diffusion of best practice and benchmarking on a number of quantifiable outputs while also trying to continuously improve process activities. A more pragmatic approach is to take a holistic approach that focuses on efficiency and effectiveness and therefore captures processes and activities and the broader contributions to business and society. For example as an indicator, the revenues received from contract research could be complemented with the average size and length of the contracts to provide a sense of the depth of the research assignments and the multiplier impact on the local, regional, or national economy. This should be the objective of any evaluation mechanism introduced to monitor the activities of TTOs.

The introduction of measurement systems generally helps to encourage behaviour by defining activities that previously went unnoticed or were unusual. Often actors will use the measurement system to guide their behaviour (Molas-Gallart et al. 2002). In this respect, measurement can form a perfect lever for the promotion, awareness, and implementation of strategic objectives (Simons 1994). Some guidelines in relation to the development and use of indicators include the following:

- *Acknowledging 'variety of excellence'.* Indicators need to be sensitive to disciplinary effects and avoid biases that may reward those disciplines that exhibit the most visible direct link to commercialisation activities or successes.
- *Commercialisation indicators are not enough.* Indicators of university commercialisation are not a sufficient guide for technology transfer activities. Commercial activities are heavily concentrated in particular disciplines and the returns to commercial activities are highly skewed. On their own, commercialisation indicators are a poor reflection of the overall economic and social benefits of the university sector.
- *Use a variety of indicators.* There are no magic bullets in indicators of commercialisation activities. A variety of indicators need to be collected.
- *Some data is better than no data at all.* Resource and time constraints on TTOs often render a detailed collection of data on indicators a futile mission. It following a rationale akin to Molas-Gallart et al. (2002) has to be appreciated that even small efforts in this area can be extremely beneficial.

When considering the potential use of new indicators, it is important to assess the effort necessary to collect, analyse, and update the data. It is common to underestimate the work that is needed to collect comprehensive data necessary to carry out proper evaluations and impact assessments. Labour-intensive techniques can be applied in one-off studies but cannot be used as the basis of continued comprehensive studies over time.

4.4.2 Continuous Review of Best Practice

It is important that research institutions, and TTOs in particular, continuously seek to operate best practice methods for increasing research commercialisation activity. This includes closely monitoring practices in other countries and should include overseas study trips to specifically review practices relating to technology transfer and research commercialisation. These reviews and fact-finding missions could be carried out on a collaborative basis. Furthermore, the horizontal collaboration between universities or meso-level structures referred in Chap. 3 should facilitate in communicating and diffusing best practices across a number of institutions. Therefore, TTO should considership membership of professional bodies such as ATUM, Praxis or ASTP that is one element of support in their consistent pursuit of improvements.

4.4.3 Annual Review/Report

It is now a commonplace that TTOs produce annual reports of their activities. Annual reports or other reports noting evaluations can also enhance the reputation and credibility of the TTO among industry. Such reports document the extent and level of technology transfer and licencing activities as well as recent policy development and changes. These reports are made available through different channels and serve as a useful tool to communicate the role and remit of the TTO to internal and external stakeholders as well as promoting the professionalism of the office. Reports also document success stories and provide details of the processes of commercialisation used in each of these cases. These accounts can be useful in altering mindsets and attitudes to commercialisation. For smaller institutions, an annual report may be too ambitious undertaking but rather the TTO should attempt to systematically review its operations and its effectiveness in carrying out its obligations to all stakeholders every 2 or 3 years. Publication of the methodology and results of this review, taking account of the stated objectives of the TTO, its performance against key performance indicators, evaluations of management and activities, should ensure transparency, and communicate the activities of the office to all stakeholders.

There is obviously a necessity and benefit for TTOs to document and report the activities it is conducting and the outcomes of such activity. Increasingly funding bodies are putting pressure on TTOs to develop such evidence on a frequent basis. Commercialisation plans and abilities are often part of funding criteria and so it is beneficial to the university that the TTO can demonstrate professionalism and proactiveness in this respect. Crucially users should be involved in the design of evaluation measures as this creates buy-in and awareness.

Another aspect of the evaluation process may involve asking industry partners for an evaluation of the TTOs services. One university studied by Scott et al. (2001), for example hired an outside firm to measure customer satisfaction with its Technology Licencing Office. The extent and operation of evaluation procedures will be contingent on each university and the available resources. Evidence suggests that even minor efforts to evaluate activity can prove beneficial—indicating to TTO staff the areas they should be emphasising, projecting an image of professionalism and transparency, and ultimately benefiting the technology transfer process.

4.5   Culture and Ethos for Commercialisation

The cultural divide between researchers and business has often been acknowledged as a major impediment to knowledge creation (Leydesdorff 2000). Technology transfer and the central objectives of technology transfer cannot take place without a supportive cultural context. While the complexities of commercialisation on the university's mission has been highlighted previously efforts should be made to support commercialisation activities for those who wish to voluntarily avail of them.

> **Key Success Factor**
> A strong university culture of research, innovation, and entrepreneurship.

In order for university staff and researchers to fully embrace the concept of commercialisation, there needs to be a change of mindset. Companies and universities are not natural partners: their cultures and their missions are different. Academics value their freedom and independence. An overarching aim for the TTO should be to take part in efforts to reduce these discrepancies by highlighting the mutual benefits to be gained from commercialisation activities and nurturing trust and common expectations between partners. As noted by Beesley (2003): 'it is apparent that a major cultural shift is necessary if collaborative networks are first to be established and second to reach their full potential'.

A cultural change will only be effective if it is introduced at all levels of institutions. Crucially, any development of consensus and emphasis on values, which appreciate commercialisation will require top-down leadership from the Dean and Head of Department (Organ and Cunningham 2011). Such visible commitment by top management to promote greater encouragement of collaboration between industry and academia is a key to success and mutual learning. The appreciation of commercial activities as complementary to the traditional teaching and research activities of universities will evolve gradually over time. It also requires that commercialisation is part of the promotion criteria for career progression. As the OECD (2003: 17) noted: 'regulations are not sufficient what is required is a gradual change of mindset and an appreciation of the value of such activities in enhancing the role of universities in society and embedding their positive effects'.

4.5.1 Consensus and Mutual Understanding

Often the potential still exists for commercialisation, but it is undermined by a lack of mutual understanding (Leydesdorff 2000). One of the challenges for TTO has been a lack of consensus within institutions with regard to the endorsement of commercialisation as a valuable outlet for research results (e.g. Mangematin et al. 2014; Siegel et al. 2003; Lambert 2003). Within universities, therefore, need to be made to develop a consensus on research commercialisation. Academics and industry need to understand the role of commercialisation, its relation to the university's mission, and its benefit for the public good. Within universities, there are ongoing debates and different opinions regarding the commercialisation of research and the role of researchers in this process. Research shows that academics have useful insights into the process, which TTOs could draw on. As part of attempts to develop a consensus as to the benefits of commercialisation stakeholders should be consulted and informed, in order to create support and buy-in.

Another mechanism to promote a change in mindset may be to locate researchers alongside technology companies. This has been found to have a positive impact on research commercialisation and also sends out an implicit message of the universities' intention. The Turku Technology Centre represents a historical example in this regard. This mechanism of co-location creates an environment with increased opportunity collaboration. Crucial in initiating cultural change initiatives is the dissemination of success stories and the benefits that have arisen from the process. Role models have been instrumental in encouraging commercialisation—both for intellectual property exploitation and for spin-off enterprises. Cases of good practice assist in generating positive images of entrepreneurship within a research environment. Success stories can be promoted through both media and workshops with talks by speakers from campus companies, etc. Moreover, there needs to be recognition that there are different types of individuals who pursue academic entrepreneurship within institutions—academic entrepreneurs and entrepreneurial academics (see Miller et al. 2018).

4.5.2 Formal Recognition

Another mechanism is the formal recognition of role models by the institution through awards such as Presidential medals or nomination for excellence in research. For example, a scheme could be established to recognise researchers who have demonstrated good practice in research commercialisation, technology transfer, and contribution to industrial development and public service generally. Such award schemes should run in parallel to existing award schemes running in most universities for excellence in teaching and research.

The potential researcher benefits of greater interaction with industry should be actively promoted. Collaborative ventures have been highlighted by researchers as a valuable means of identifying research ideas with commercial or industrial applications. The level of technology transfer has increased in environments where there is greater interaction with industry on an ongoing basis. There are a number of mechanisms that can be used to increase such interaction, including the use of industry advisory groups to provide advice and feedback to individual research groups.

In a broader university context, the creation of an entrepreneurial climate can be linked to the curricula in place at both undergraduate and postgraduate levels. This could include general entrepreneurship and innovation management training. Industrial application and research commercialisation could be included in undergraduate and postgraduate programmes or be part of induction sessions. It is of particular importance for science students to develop entrepreneurial skills to allow them to exploit their innovations and develop the commercial potential of their work. At the very minimum, they need to have an understanding of new venture creation and different aspects of entrepreneurship and innovation. Such offering may not be explicitly called entrepreneurship and innovation so as to ensure that participants are not turned off by such terminology. Universities provide different institutional supports to underpin such activities, which can be housed in universities' entrepreneurship centres (see Dolan et al. 2019)

A supportive cultural context is fundamental to technology transfer initiatives. Technology transfer occurs when university faculty and industry work together for mutual gain. The industry can gain significant benefits from their collaboration with universities (see Cunningham and Link 2015). Therefore, industry–university collaboration cannot be forced and cultural differences must be understood and attempt made at consensus and understanding (Fassin 2000: 33). One of the greatest challenges in terms of developing TTOs will be a change of mindset in terms of the appropriateness and importance of commercialisation-related activities in the context of the university's overall mission. We have drawn together factors that may facilitate in altering mindsets and value appreciation (see Table 4.2).

By developing awareness, culture, and ethos for commercialisation the academic researcher will be more willing to undertake research commercialisation. Key facilitating factors include the establishment of a consensus approach to research commercialisation, the provision of information, and dissemination and publicity of

Table 4.2 Factors facilitating a changing in mindsets

- Establish an atmosphere that promotes research among faculty
- Foster/develop/encourage a cooperative mindset among researchers and the TTO
- Leadership from the top levels
- Encourage research directors to promote awareness of technology transfer among researchers
- Make the Technology Transfer Office visible to researchers
- Develop technology transfer policies via a participatory process (i.e. use faculty input to draft policies)
- Develop policies and procedures that are easy to understand
- Interact with researchers on a one to one basis
- Develop a royalty structure that allows the inventor some flexibility
- Cultivate university/industry peer relationships with networking opportunities
- Keep the inventor informed and in the loop during commercialisation activities
- Guide the inventor: Be willing to discuss whether it is worth carrying research further for commercialization purposes
- Share successful cases with researchers
- Clearly relate activities to the university mission

role models and research commercialisation success stories to both staff and students to motivate future creative innovators. Further, interrelated issues with regard to the motivation of researchers are discussed in the following section.

4.6   Researcher Motivation

Key Success Factor
Researcher Motivation and Commitment.

4.6.1   Personal Motivation

The most common stimulant to commercialisation of research as identified by the majority of research is the personal motivation of the individual researcher or academic. Invention disclosures are the crucial input into initiating the technology transfer process and cannot be simply extracted from researchers but rather must be submitted on a voluntary basis. This personal motivation often stems from the background of the researcher in terms of their training and professional experience. Researchers who have previous industry experience will find it easier to collaborate and work with industry and there are some gender differences in this regard (Cunningham et al. 2017). An important element of the personal motivation of researchers to commercialise their research is seeing a product or service derived from their knowledge discovery that is available to the public or is adopted by industry.

Researcher motivation is crucial as the number of disclosures will depend, to some extent, on the efforts of the TTO to elicit disclosures and faculty interest in technology transfer (O'Reilly and Cunningham 2017; O'Kane et al. 2017; Siegel et al. 2003). Evidence shows that there is considerable heterogeneity across universities in their ability to elicit invention disclosures from their faculty. Interacting with the institution's TTOs takes time and may not have obvious benefits for a faculty member. There is some evidence that a significant number of inventions still go unreported and unpatented (Hall 2004). A number of activities can be seen to provide researcher motivation and facilitate in increasing the number of inventions disclosed by researchers, as discussed below.

4.6.2 Executive Management

Within universities, it is important to have well-defined lines of responsibility, clearly delegated authority, and cohesive management teams of academics and administrators. The delineation of institutional responsibilities needs to be clear for internal and external stakeholders. So for example which institutional portfolio does the TTO reside and how is in represented within a university executive management team? Moreover, given the dynamic environment of TTOs and the need for effective decision-making then different institutional mechanisms need to be put in place to facilitate and support such activities. The support provided by such executive management can help promote the TTO as a professional and efficient mechanism to researchers for the exploitation of their research.

4.6.3 Commercialisation Knowledge and Skills

Sometimes researchers are motivated to pursue commercialisation activities simply because they are unaware of the benefits that exist or the process of commercialisation broadly.

> **Key Success Factor**
> HR strategies that promote, recognise, and reward commercialisation activities.

Successful technology transfer depends, above all, on the interest and enthusiasm that faculty scientists bring to the joint research effort. Promoting such motivation generally falls under the auspices of general university policy. Thus, university policy can be indicative of the level of support and recognition for commercialisation activities generally (Scott et al. 2001). The development and implementation of successful human resource strategies are among the most important tasks facing universities. Human resource strategies should promote and support the desired employee role behaviours among university staff. While human resource strategies used to support commercialisation activities cannot be introduced in all institutions immediately, they can be gradually introduced in recognition of the increasing importance of such activities. The integration of commercialisation activities into human resource strategies may become manifest by for example allowing more leave of absence for staff, reduced workloads or simply in terms of professional development. Human resource management and other organisational practices that influence such incentives could explain some of the variations in technology transfer performance across universities (Siegel et al. 2003). The main human resource strategy levers, which may provide motivation to researchers include orientation programmes, reward mechanisms, and sabbaticals for researchers in industry.

4.6.4 Orientation Programmes

An introductory and extremely necessary step for successful development of technology transfer is that components of induction models should include briefings on commercialisation of research and management of IP. Indeed, in most universities, it is the policy that a condition of taking up employment is that new staff members participate in an orientation programme, which includes basic training modules in intellectual property rights, technology transfer, and the institution's support structures for academic entrepreneurship. Induction modules also provide an opportunity for communicating how such activities link with the objectives of the universities' overall strategic plan. While communication with incoming staff will generate a collective understanding training modules could also be set up to educate existing staff as to the benefits of commercialisation and university policies in place. Permanent senior researchers often undertake a more intensive programme.

4.6.5 Reward Mechanisms

Generally, the current merit system in research institutions does not explicitly value commercialisation. As is evidenced from research in the US rules on ownership, in themselves are not sufficient, there must be incentives for institutions and for researchers to protect and exploit IP resulting from public research (OECD 2004). In order to encourage employees and increase motivation university reward systems could be modified to be consistent with technology transfer objectives. Siegel et al. (2003) note that: 'the propensity of faculty members to disclose inventions, and thus increase the supply of technologies available for commercialisation will be related to promotion and tenure policies and the university's royalty and equity distribution formula'. Typically, researchers involved in commercialisation activities are financially rewarded through a scheme operated by the TTO such percentage of income generated from royalties and licences. Most universities have developed a formulae for this, which reflects their underlying values and priorities. Many institutions allow research staff to generate considerable income. In many cases, such income is also reverted to an institutional or departmental development fund, to support ongoing research. It is reinvested by faculty and the university to support research. This, in turn, creates a strong incentive for research teams since it produces new opportunities for further research work (Hernes and Martin 2000). There is a widespread belief that there are insufficient rewards for faculty involvement in technology transfer (Allan 2001). There is also hesitancy and resistance to suggested change in the human resource strategies of universities and the behaviours they seek to support. These stem from fears that nurturing commercialisation activities may divert the direction of enquiry and draw attention away from more traditional university activities. Incentives for technology transfer should, therefore, be carefully promoted as a complement to existing policies. Universities must become more flexible in their reward systems for those who choose to engage in collaborative arrangements of scientific enquiry (Beesley 2003: 1529). The strategic objectives of the university usually suggest support for these types of motivating mechanisms.

Researchers who engage in commercialisation activities should, therefore, be afforded the same prestige through collaborative and transdisciplinary work as they do within their existing disciplinary structures. Current research structures and environments have given rise to the idiom 'publish or perish' and are in direct conflict with collaborative research. Reward systems, if universities are to fully embrace of trilateral arrangements of science, need to focus on the rigor and of outcomes, rather than the field of research or the ability to publish. Often reward structures are inconsistent with the organisational objective of increasing the role of technology transfer as an inherent part of accomplishing the university mission. Well-managed staff promotion procedures that reward success in research are therefore important stimuli to good and sustained performance, just as poorly managed or arbitrary procedures that seem to prefer age and seniority, to talent and success, are important demotivators (Shattock 2001).

Hiring, tenure, and promotion processes should give appropriate credit to university researchers who collaborate with industry. Institutions should, therefore, review and examine incentives and rewards provided to faculty and support staff for participating in technology transfer. Some universities have even introduced incentives specifically for University Technology Managers, although often university or national policies can constrain incentives in this area unless the TTO is set up as a separate legal entity. Evaluation of commercialisation for a merit system should include activities such as promoting research partnerships, the application of research results and technology transfer.

Further reward mechanisms operating in other HEIs include sharing of the campus portion of the licence income, for education, and research purposes, at a broad range of levels. Intrinsic recognition in the form of presentation of awards honouring inventors and entrepreneurs can serve as excellent motivating factors. The purpose of such human resource strategy initiatives is to institutionalise a commercial ethos. Caution must be exercised, however, that incentive systems do not encourage actors to accumulate solely quantifiable activities to the neglect of quality and value (OECD 2003: 14).

Ultimately, research collaborations must be based on the willingness and enthusiastic participation of individual faculty members. Researchers should not be penalised for taking the commercialisation route but rather support should permeate into the human resource strategies of universities. The more progressive steps of redefining scholarship to include technology transfer or considering participation in commercialisation as an element in tenure or promotion decisions are necessary. For most universities, a more realistic introductory step may be to ensure that support for commercialisation activity is clearly stated. The critical importance of this is emphasised by research by Siegel et al. (2003), which concluded that: 'the most critical organisational factors influencing technology transfer are likely to be reward systems for faculty, TTO staffing, and compensation practices as well as actions taken by administrators to reduce informational and cultural barriers between universities and firms'.

4.6.6   Sabbaticals for Researchers in Industry

Research institutions should actively encourage researchers to take sabbatical leave to work in an industry for a specified time. The terms and conditions of sabbatical leave should be flexible enough to foster academic entrepreneurship. This tends to facilitate technology transfer to the company via the researcher's know-how. In return, the institution will enhance its linkages with the company and the researcher will develop better insights into the industry's requirements and practical aspects of achieving technology transfer. Other flexible terms of employment may also be important.

4.6.7   Develop Full-Time Research Career Opportunities

Another factor that can encourage research commercialisation activities is the opportunity for permanent full-time research career paths for postdoctoral researchers who wish to stay in an academic research environment, but who do not wish to pursue a traditional academic career, which normally involves teaching. Research suggests that it is these types of academics, given adequate time following on from postdoctoral research that is perhaps most predisposed to commercialisation activities and developing linkages with industry (Graff et al. 2002).

4.7   Conclusions: Lessons for University Leaders and Managers

The lessons in terms of developing a culture conducive to technology transfer that motivates and provides incentives for researchers. In our business model framework, we have captured core issues central to developing a strategic approach to technology transfer. Taking the key factors for success and lessons that can be drawn from the discussion in each of these issues are outlined in Table 4.3 that are aligned to our business model framework presented in Chap. 3. Our key message is that universities need to take a strategic approach to technology transfer and be realistic and pragmatic. TTOs take time to evolve and flourish and their success is dependent on a range of factors much of which is outside of their direct control. Effective TTOs can provide universities with many tangible and intangible benefits, but more importantly they can facilitate and support the university engagement in different spheres within region, national, and international environments.

Table 4.3 Business model framework: key factors for sucess and facilitating factors

Key issues	Key factors for success	Facilitating factors
Strategic technology transfer: Key issues		
1. Strategic management of technology transfer	Development of a mission statement, strategic plan, and strategic priorities	• Link objectives with the university's strategic plan • Demonstrate how activities contribute to achieving aspects of the universities mission • Top leadership commitment • Widespread communication of objectives • Stakeholder input into objectives • Acceptance and agreement of objectives by multiple stakeholders • Mission statement that refers to the broad benefits of commercialisation for the public good • Title of office that reflects the chosen strategic priorities • Executive level committees
2. Technology transfer activities	Focus on action as opposed to facilitation	• Assess and prioritise activities in relation to strategic objectives • Build on clusters of knowledge
	Identification of technology platforms	• Communicate elements of focus to business • Target research funding • Actively monitor research activities
	Active networking	• Allocate time for attendance at conferences, etc. • Databases of contracts and research

(continued)

Table 4.3 (continued)

Key issues	Key factors for success	Facilitating factors
		• Horizontal networking with other Institutions • Membership of associations/societies
	Focus on core activities and development of core competencies	• Shed non-core activities • Focus on getting the deal done
	Boundary spanning	• Relationship management • Clear communication • Database to track/manage inventions • Outreach and education
3. Structuring technology transfer	There is no one best way to structure activities	• Ensure clear lines of reporting and responsibility • TTO must have a prominent position in university structure and among top management • Improve communication and teamwork among university personnel • Co-ordinate the efforts of various offices to support researchers • Close interaction between commercialisation and research activity • Typically, TTO Directors report to Vice President/Vice Rector for Research • Matrix Structures best for co-ordinating and incentivising university-wide interface with business • Although research and technology transfer are related the two offices should be administratively independent • Executive Committee to oversee commercialisation activities • Co-location of offices to encourage teamwork • Structure by Specialism (Activity or Case Manager in specific research area)
	Shared services Meso-level structures to facilitate co-operation	• Embrace innovative and creative mechanisms to overcome deficiencies • Complement on the ground presence with shared services to overcome resource deficiency (especially related to patenting and legislative issues) • Shared Services could include licencing negotiation market research for new technologies, IP management,

(continued)

Table 4.3 (continued)

Key issues	Key factors for success	Facilitating factors
		Legal Expertise, Spin-out creation • university-based services could include raising awareness of IP, negotiating collaborative research contracts, local contact • Initiatives should be supported financially by government agencies. Agencies may also provide expertise to help manage and co-ordinate the structure • Volunteer TTOs in each region should agree themselves how to set up and shape the services, and the role that each institution should play • Strict policies regarding the conduct of member institutions particularly in terms the ownership of Intellectual Property and the confidentiality of information
4. Staffing, skills, and resources	Strategic staffing and continuous training and development	• Need close links with researchers—facilitated by academic credentials • Diverse skill base of TTO staff—understand science and willingness to put a business case around it • Ability to make 'cold calls' • Good people management skills • Leadership is crucial • Accreditation and professional development standards • Case managers should be trained in a particular area of science and trained in IP management
	Integrate HR strategy of TTO with strategic objectives	• Policies that nurture and reward commercialisation activities • Performance appraisal of staff against objectives • Induction policy • Placement of staff/Internship in other Technology Transfer Offices
5. Policies and procedures	Requirement for user friendly and clearly documented IP policies	• Inclusion of policies on Video, CD Rom, or pocket-sized summaries • Researcher kits for each agreement • Advisory committee made up of stakeholders to facilitate development • Timely responses to invention disclosures • Communicating clearly with researchers • Use of templates/boilerplates or master contracts to negotiate agreements

(continued)

Table 4.3   (continued)

Key issues	Key factors for success	Facilitating factors
		• Publication delays carefully monitored to preserve academic freedom
	Intellectual property rights	• Clear ownership policies • Policies for the distribution of royalties, preferably weighted positively towards the inventor • Harmonisation in terms of national policy on IP • Flexibility in the application of IP rights between universities and business • Protocol for research collaborations • universities should not be overly aggressive exercising university IP rights • Policies for resolving conflicts of interest • Keep legal fees in check
6. Mechanisms for technology transfer	Spin-offs and start-ups	• Provide clear and transparent guidelines • Proactively provide information, contacts, and support • Jointly discuss IP and decide at the outset with the academic inventors or prospective entrepreneurs whether setting up a company is the best option • Prospective entrepreneurs should be allowed to remain on a part-time research or teaching position with the universities if they so desire
	Risk of spin-outs	• Danger of too many unsustainable low-quality spin-outs • Need to assess the likelihood of success, the resource requirement, and the timeline • Develop expertise in this area or else the activity becomes unsustainable and a higher risk for the university
	Incentive structure for spin-outs	• university managed incubator facilities • Provision of seed funding in return for equity • Focused efforts to attract management talent and catalyse company formation

(continued)

Table 4.3 (continued)

Key issues	Key factors for success	Facilitating factors
	Training/seminars for prospective academic entrepreneurship	• Training courses and information seminars in enterprise development for researchers
	licencing	• IP should be taken out selectively • Develop capacity and resources for managing IP • Proof-of-concept funding (average level of investment is up to 70,000 € per invention) in conjunction with development agencies and other TTOs
	Portfolio approach: pursuing many deals	• TTO pursues a number of commercialisation channels to minimise the risk exposure • AUTM established a positive correlation between royalty income and the number of active licences at an academic institution • Need to prioritise activities (1 in 9 deals pays out royalties) • More realistic in a resource constraint environment
	Networking	• To evaluate disclosures • Facilitating commercialisation opportunities • Formal and informal
7. Evaluation procedures	Develop a holistic approach to evaluating technology transfer activities	• Not merely performance metrics but promoted for the way they facilitate to support the management of activities • Focus on commercialisation outputs and TTO processes and activities • Use of a variety of indicators • Deal flow is critical • Link indicators to strategic priorities • Top management support • User involvement in the development of metrics • Acknowledge 'variety of excellence' • Sufficient resources allocated to collection and analysing • Continuous review of best practice • Annual review/report to communicate results
Key issues in developing a supportive culture		
8. Culture and ethos for commercialisation	A strong university culture of research, innovation, and entrepreneurship	• Highlight mutual benefits to be gained from commercialisation activities • Clearly relate activities to the university mission • Top level support and commitment to commercialisation

(continued)

Table 4.3 (continued)

Key issues	Key factors for success	Facilitating factors
		• Develop a consensus on research commercialisation • Develop policies via a participatory process • Locate researchers alongside technology companies • Promote and communicate success stories • Formal recognition for commercialisation activities • Entrepreneurship and innovation management training • Undergraduate/postgraduate curricula including entrepreneurship, etc. • Make the ILO visible to researchers
9. Researcher motivation	Motivated and entrepreneurial researchers	• Keep the inventor informed and in the loop during commercialisation activities • User-friendly policies • Relationship management • Clear communication • Database to track/manage Inventions • Outreach and education
	HR strategies that recognise and reward commercialisation	• Orientation programmes with IP modules • Reward mechanisms that recognize commercialisation activities • Inclusion of commercialisation activities in tenure and promotion decisions • Sabbaticals for researchers in industry • Full-time research opportunities • Permanent/Contract ratio alignment • Incentives and flexibility in employment conditions to stimulate spin-off activity

References

Allan MF (2001) A review of best practices in university technology licensing offices. J Assoc Univ Technol Manager 13(1):57–69

Association of University Technology Managers (2002) Licensing survey FY 2002 survey summary. Northbook, IL

Beesley LG (2003) Science policy in changing times: are governments poised to take full advantage of an institution in transition? Res Pol 32(8):1519–1531

Benneworth P, Charles D (2005) University spin-off policies and economic development in less successful regions: learning from two decades of policy practice. Eur Plann Stud 13(4):537–557

Bergek A, Norrman C (2008) Incubator best practice: a framework. Technovation 28(1–2):20–28

Binsawad M, Sohaib O, Hawryszkiewycz I (2019) Factors impacting technology business incubator performance. Int J Innovat Manag 23(01):1950007

Cunningham JA, Link AN (2015) Fostering university-industry R&D collaborations in European Union countries. Int Entrepren Manag J 11(4):849–860

Cunningham J, O'Reilly P, O'Kane C, Mangematin V (2014) The inhibiting factors that principal investigators experience in leading publicly funded research. J Technol Transf 39(1):93–110

Cunningham JA, O'Reilly P, Dolan B, O'Kane C, Mangematin V (2017) Gender differences and academic entrepreneurship: A study of scientists in the principal investigator role. In: Link AN (ed) Gender and entrepreneurial activity. Edward Elgar, Cheltenham

Cunningham JA, Romano M, Nicotra M (2020) A European perspective on intellectual property and technology transfer. In: Research handbook on intellectual property and technology transfer. Edward Elgar Publishing

Dolan B, Cunningham JA, Menter M, McGregor C (2019) The role and function of cooperative research centers in entrepreneurial universities: A micro level perspective. Manag Decis. https://doi.org/10.1108/MD-10-2018-1172

Fassin Y (2000) The strategic role of university-industry liaison offices. J Res Admin 1(2):31–42

Graff G, Heiman A, Zilberman D, Castillo F, Parker D (2002) Universities, technology transfer and industrial R&D. In: Evenson RE, Santaniello V, Zilberman D (eds) Economic and social issues in agricultural biotechnology. CAB, Wallingford, pp 93–117

Hall, BH (2004) University-industry research partnerships in the United State. Conference paper presented at Kansia Conference

Hernes G, Martin M (2000) Management of university-industry linkages. Policy forum no. 11, UNESCO/International Institute for Educational Planning, Paris. http://unesdoc.unesco.org/images/0012/001235/123538e.pdf

Kim J, Daim T, Lavoie JR (2019) Technology licensing performance and strategy of US research institutions. In: Daim TU, Dabić M, Başoğlu N, Lavoie JR, Galli BJ (eds) R&D management in the knowledge era. Springer, Cham, pp 531–549

Lambert R (2003) Lambert review of business-university collaboration. Final report, December

Leydesdorff L (2000) The triple helix: an evolutionary model of innovations. Res Pol 29(2):243–255

Link AN, Scott JT (2017) Opening the ivory tower's door: An analysis of the determinants of the formation of US university spin-off companies. In: Universities and the entrepreneurial ecosystem. Edward Elgar, Cheltenham

Mangematin V, O'Reilly P, Cunningham J (2014) PIs as boundary spanners, science and market shapers. J Technol Transf 39(1):1–10

Mathieu G (2012) Conflict of interest in university technology transfer

Meyer JF (2003) Strengthening the technology capabilities of Louisiana universities. Pappas & Associates Report, Prepared for the Louisiana Department of economic Development, 30 June 2003

Miller K, Alexander AT, Cunningham J, Albats E (2018) Entrepreneurial academics and academic entrepreneurs: A systematic literature review. Int J Technol Manag 77(1–3):9–37

Miranda FJ, Chamorro A, Rubio S (2018) Re-thinking university spin-off: A critical literature review and a research agenda. J Technol Transf 43(4):1007–1038

Molas-Gallart J, Salter A, Patel P, Scott A, Duran X (2002) Measuring third stream activities. Final report to the Russell Group of Universities, SPRU, Science and Technology Policy Research, April

Mustar P, Wright M, Clarysse B (2008) University spin-off firms: lessons from ten years of experience in Europe. Sci Public Pol 35(2):67–80

OECD (2002) Interim results of the TIP project on the strategic use of IPRs at PROs, Paris

OECD (2003) Report on OCED-China events on intellectual property rights held in Bejing, China. httpe.www.oecd.org/dataoecd/46/10/32267101.pdf

OECD (2004) Report on OECD-China Events on Intellectual Property Rights held In Beijing, China

O'Kane C, Zhang JA, Cunningham JA, O'Reilly P (2017) What factors inhibit publicly funded principal investigators' commercialization activities? Small Enterprise Res 24(3):215–232

O'Reilly P, Cunningham JA (2017) Enablers and barriers to university technology transfer engagements with small-and medium-sized enterprises: perspectives of principal investigators. Small Enterp Res 24(3):274–289

Organ DJ, Cunningham J (2011) Entrepreneurship in the Academy: The case for a micro-institutional analysis. DRUID 2011 on innovation, strategy, and structure—organizations, institutions, systems and regions at Copenhagen Business School, Denmark, June 15–17

Rasmussen E, Mosey S, Wright M (2011) The evolution of entrepreneurial competencies: A longitudinal study of university spin-off venture emergence. J Manag Stud 48(6):1314–1345

Rothaermel FT, Thursby M (2005) University–incubator firm knowledge flows: assessing their impact on incubator firm performance. Res Pol 34(3):305–320

Sanchez AM, Tejedor ACP (1995) University-industry relationships in peripheral regions: The case of Aragon in Spain. Technovation 15(10):613–625

Schwartz M, Hornych C (2010) Cooperation patterns of incubator firms and the impact of incubator specialization: Empirical evidence from Germany. Technovation 30(9–10):485–495

Scott A, Steyn G, Geuna A, Brusoni S, Steinmueller E (2001) The economic returns to basic research and the benefits of university-industry relationships. A literature review and update of findings. Report for the Office of Science and Technology by SPRU

Shattock ML (2001) In what way do changing university-industry relations affect the management of higher education institutions, Part III. In: Hernes G, Martin M (eds) Management of university-industry linkages, Policy Forum No.11, UNESCO/International Institute for Educational Planning, Paris

Siegel, D. S., Waldman, D., & Link, A. (2003). Assessing the impact of organizational practices on the relative productivity of university technology transfer offices: an exploratory study. Research policy, 32(1), 27–48

Simons R (1994) How new top managers use control systems as levers of strategic renewal. Strateg Manag J 15(3):169–189

Thompson NC, Ziedonis AA, Mowery DC (2018) University licensing and the flow of scientific knowledge. Res Pol 47(6):1060–1069

Thursby JG, Thursby MC (2002) Who is selling the ivory tower? Sources of growth in university licensing. Manag Sci 48(1):90–104

Thursby JG, Thursby MC (2007) University licensing. Oxf Rev Econ Pol 23(4):620–639

University-Industry Collaboration Initiative (RCI) (2003) Working together, creating knowledge. Business Higher Education Forum

Printed in the United States
By Bookmasters